JOURNAL OF APPLIED LOGICS - IFCOLOG JOURNAL OF LOGICS AND THEIR APPLICATIONS

Volume 6, Number 4

June 2019

Disclaimer

Statements of fact and opinion in the articles in Journal of Applied Logics - IfCoLog Journal of Logics and their Applications (JALs-FLAP) are those of the respective authors and contributors and not of the JALs-FLAP. Neither College Publications nor the JALs-FLAP make any representation, express or implied, in respect of the accuracy of the material in this journal and cannot accept any legal responsibility or liability for any errors or omissions that may be made. The reader should make his/her own evaluation as to the appropriateness or otherwise of any experimental technique described.

ISBN 978-1-84890-306-7
ISSN (E) 2631-9829
ISSN (P) 2631-9810

College Publications
Scientific Director: Dov Gabbay
Managing Director: Jane Spurr

http://www.collegepublications.co.uk

EDITORIAL BOARD

SCOPE AND SUBMISSIONS

This journal considers submission in all areas of pure and applied logic, including:

pure logical systems
proof theory
constructive logic
categorical logic
modal and temporal logic
model theory
recursion theory
type theory
nominal theory
nonclassical logics
nonmonotonic logic
numerical and uncertainty reasoning
logic and AI
foundations of logic programming
belief change/revision
systems of knowledge and belief
logics and semantics of programming
specification and verification
agent theory
databases

dynamic logic
quantum logic
algebraic logic
logic and cognition
probabilistic logic
logic and networks
neuro-logical systems
complexity
argumentation theory
logic and computation
logic and language
logic engineering
knowledge-based systems
automated reasoning
knowledge representation
logic in hardware and VLSI
natural language
concurrent computation
planning

This journal will also consider papers on the application of logic in other subject areas: philosophy, cognitive science, physics etc. provided they have some formal content.

Submissions should be sent to Jane Spurr (jane.spurr@kcl.ac.uk) as a pdf file, preferably compiled in LaTeX using the IFCoLog class file.

CONTENTS

ARTICLES

vii

EDITORIAL

ARTUR D'AVILA GARCEZ

TAREK R. BESOLD

This special issue of the Journal of Applied Logics contains original contributions which have been presented at the Thirteenth International Workshop on Neural-Symbolic Learning and Reasoning (NeSy'18), part of Human-Level Artificial Intelligence, HLAI 2018, which took place in Prague, CZ, in August 23-24, 2018. All papers included here are extended versions of workshop papers which were revised thoroughly according to the rules of the Journal.

The NeSy workshop series started at the International Joint Conference on Artificial Intelligence, IJCAI 2005 (please visit www.neural-symbolic.org for more information about the workshop series). NeSy seeks to integrate well-founded symbolic Artificial Intelligence and Computer Science Logic with robust Neural Network learning systems. It is intended to create an atmosphere of exchange of ideas, providing a forum for the presentation and discussion of the following key topics of AI today: the representation of symbolic knowledge by deep networks and connectionist systems in general, semi-supervised and relational learning in neural networks, reasoning and logical inference using recurrent networks, distilling and knowledge extraction from neural networks, integrated and hybrid neural-symbolic AI approaches, integration of logic and probabilities in neural networks, knowledge-based transfer learning using neural networks, verification of neural networks and reasoning about time, deep symbolic reinforcement learning and planning, biologically-inspired neural-symbolic architectures, neural-symbolic cognitive models and systems, and applications in robotics, simulation, fraud prevention, language processing, semantic web, software engineering, fault diagnosis, bioinformatics and visual intelligence.

In *Neural-Symbolic Computing: An effective methodology for principled integration of machine learning and reasoning*, Artur d'Avila Garcez, Marco Gori, Luis Lamb, Luciano Serafini, Michael Spranger and Son Tran revisit some of the early work on the integration of neural and symbolic approaches before placing the more recent work on end-to-end reasoning and learning in context within neural-symbolic integration. In *Semi-supervised learning using differentiable reasoning*, Emile van Krieken, Erman Acar and Frank van Harmelen introduce a differentiable reasoning approach to improve the performance of semi-supervised learning using relational background knowledge. The approach is based on Real Logic and when applied to semantic image interpretation, it highlights the difficulty of neural networks at learning contrapositives. In *High-order networks that learn to satisfy logic constraints*, Gadi Pinkas and Shimon Cohen show that symmetric networks can learn in unsupervised fashion to solve planning problems described in first-order logic. Higher-order networks and sigma-pi networks are also investigated in the context of the satisfiability problem. In *Learning representation of relation dynamics with delays and refining with prior knowledge*, Yin Jun Phua, Tony Ribeiro and Katsumi Inoue study relational learning using recurrent networks in the presence of insufficient data as part of an application in biology inspired by earlier work in Inductive Logic Programming and Statistical Relational Learning. In *Compositionality for recursive nural networks*, Martha Lewis investigates the relationship between recursive neural tensor networks and category theory with examples from language modelling. The paper also suggests a number of lines of research taking advantage of the proposed categorical compositionality of embeddings. In *Towards fuzzy neural conceptors*, Till Mossakowski, Razvan Diaconescu and Martin Glauer investigate fuzzy conceptors. Conceptors describe the dynamics of recurrent neural networks, with conceptor logic forming the basis of a class of neural-symbolic approaches. The paper introduces a fuzzy subconceptor relation and a fuzzy logic for conceptors. In *Modelling identity rules with neural networks*, Tillman Weyde and Radha Manisha Kopparti revisit Gary Marcus's identity rules to show that recurrent networks including GRUs and LSTMs fail to learn basic sequence patterns requiring the notion of equality. By contrast, when constrained with an appropriate pre-defined structure to start with, simple networks can achieve perfect recall on the same tasks.

London and Barcelona, May 2019
Artur d'Avila Garcez and Tarek R. Besold

Received May 2019

Neural-Symbolic Computing: An Effective Methodology for Principled Integration of Machine Learning and Reasoning

Artur d'Avila Garcez
City, University of London
a.garcez@city.ac.uk

Marco Gori
University of Siena
mxgori@gmail.com

Luis C. Lamb
Universidade Federal do Rio Grande do Sul
luislamb@acm.org

Luciano Serafini
Fondazione Bruno Kessler
serafini@fbk.eu

Michael Spranger
Sony Japan
michael.spranger@gmail.com

Son N. Tran*
University of Tasmania
sn.tran@utas.edu.au

We thank Richard Evans for his valuable comments and suggestions.
*Corresponding author. Authors are in alphabetical order.

Abstract

Current advances in Artificial Intelligence and machine learning in general, and deep learning in particular have reached unprecedented impact not only across research communities, but also over popular media channels. However, concerns about interpretability and accountability of AI have been raised by influential thinkers. In spite of the recent impact of AI, several works have identified the need for principled knowledge representation and reasoning mechanisms integrated with deep learning-based systems to provide sound and explainable models for such systems. Neural-symbolic computing aims at integrating, as foreseen by Valiant, two most fundamental cognitive abilities: the ability to learn from the environment, and the ability to reason from what has been learned. Neural-symbolic computing has been an active topic of research for many years, reconciling the advantages of robust learning in neural networks and reasoning and interpretability of symbolic representation. In this paper, we survey recent accomplishments of neural-symbolic computing as a principled methodology for integrated machine learning and reasoning. We illustrate the effectiveness of the approach by outlining the main characteristics of the methodology: principled integration of neural learning with symbolic knowledge representation and reasoning allowing for the construction of explainable AI systems. The insights provided by neural-symbolic computing shed new light on the increasingly prominent need for interpretable and accountable AI systems.

1   Introduction

Current advances in Artificial Intelligence (AI) and machine learning in general, and deep learning in particular have reached unprecedented impact not only within the academic and industrial research communities, but also among popular media channels. Deep learning researchers have achieved groundbreaking results and built AI systems that have in effect rendered new paradigms in areas such as computer vision, game playing, and natural language processing [27, 45]. Nonetheless, the impact of deep learning has been so remarkable that leading entrepreneurs such as Elon Musk and Bill Gates, and outstanding scientists such as Stephen Hawking have voiced strong concerns about AI's accountability, impact on humanity and even on the future of the planet [40].

Against this backdrop, researchers have recognised the need for offering a better understanding of the underlying principles of AI systems, in particular those based on machine learning, aiming at establishing solid foundations for the field. In this respect, Turing Award Winner Leslie Valiant had already pointed out that one of the key challenges for AI in the coming decades is the development of integrated reasoning and learning mechanisms, so as to construct a rich semantics of intelligent cognitive behavior [54]. In Valiant's words: *"The aim here is to identify a way of looking at and manipulating commonsense knowledge that is consistent with and can support what we consider to be the two most fundamental aspects of intelligent cognitive behavior: the ability to learn from experience, and the ability to reason from what has been learned. We are therefore seeking a semantics of knowledge that can computationally support the basic phenomena of intelligent behavior."* In order to respond to these scientific, technological and societal challenges which demand reliable, accountable and explainable AI systems and tools, the integration of cognitive abilities ought to be carried out in a principled way.

Neural-symbolic computing aims at integrating, as put forward by Valiant, two most fundamental cognitive abilities: the ability to learn from experience, and the ability to reason from what has been learned [2, 12, 16]. The integration of learning and reasoning through neural-symbolic computing has been an active branch of AI research for several years [14, 16, 17, 21, 25, 42, 53]. Neural-symbolic computing aims at reconciling the dominating symbolic and connectionist paradigms of AI under a principled foundation. In neural-symbolic computing, knowledge is represented in symbolic form, whereas learning and reasoning are computed by a neural network. Thus, the underlying characteristics of neural-symbolic computing allow the principled combination of robust learning and efficient inference in neural networks, along with interpretability offered by symbolic knowledge extraction and reasoning with logical systems.

Importantly, as AI systems started to outperform humans in certain tasks [45], several ethical and societal concerns were raised [40]. Therefore, the interpretability and explainability of AI systems become crucial alongside their accountability.

In this paper, we survey the principles of neural-symbolic integration by highlighting key characteristics that underline this research paradigm. Despite their differences, both the symbolic and connectionist paradigms, share common characteristics offering benefits when integrated in a principled way (see e.g. [8, 16, 46, 53]). For instance, neural learning and inference under uncertainty may address the brittleness of symbolic systems. On the other hand, symbolism provides additional knowledge for learning which may e.g. ameliorate neural network's well-known catastrophic forgetting or difficulty with extrapolating. In addition, the integration of neural models with logic-based symbolic models provides an AI system capable of bridging lower-level information processing (for perception and pattern recognition) and higher-level abstract knowledge (for reasoning and explanation).

In what follows, we review the important and recent developments of research on neural-symbolic systems. We start by outlining the main important characteristics of a neural-symbolic system: Representation, Extraction, Reasoning and Learning [2, 17], and their applications. We then discuss and categorise the approaches to representing symbolic knowledge in neural-symbolic systems into three main groups: rule-based, formula-based and embedding-based. After that, we show the capabilities and applications of neural-symbolic systems for learning, reasoning, and explainability. Towards the end of the paper we outline recent trends and identify a few challenges for neural-symbolic computing research.

2   Prolegomenon to Neural-Symbolic Computing

Neural-symbolic systems have been applied successfully to several fields, including data science, ontology learning, training and assessment in simulators, and models of cognitive learning and reasoning [5, 14, 16, 34]. However, the recent impact of deep learning in vision and language processing and the growing complexity of (autonomous) AI systems demand improved explainability and accountability. In neural-symbolic computing, learning, reasoning and knowledge extraction are combined. Neural-symbolic systems are modular and seek to have the property of compositionality. This is achieved through the streamlined representation of several knowledge representation languages which are computed by connectionist models. The *Knowledge-Based Artificial Neural Network* (KBANN) [49] and the *Connectionist inductive learning and logic programming* (CILP) [17] systems were some of the most influential models that combine logical reasoning and neural learning. As pointed out in [17] KBANN served as inspiration in the construction of the CILP system. CILP provides a sound theoretical foundation to inductive learning and reasoning in artificial neural networks through theorems showing how logic programming can be a knowledge representation language for neural networks. The KBANN system was the first to allow for learning with background knowledge in neural networks and knowledge extraction, with relevant applications in bioinformatics. CILP allowed for the integration of learning, reasoning and knowledge extraction in recurrent networks. An important result of CILP was to show how neural networks endowed with semi-linear neurons approximate the fixed-point operator of propositional logic programs with negation. This result allowed applications of reasoning and learning using backpropagation and logic programs as background knowledge [17].

Notwithstanding, the need for richer cognitive models soon demanded the representation and learning of other forms of reasoning, such as temporal reasoning, reasoning about uncertainty, epistemic, constructive and argumentative reasoning [16, 54]. Modal and temporal logic have achieved first class status in the formal toolboxes of AI and Computer Science researchers. In AI, modal logics are amongst the most widely used logics in the analysis and modelling of reasoning in distributed multiagent systems. In the early 2000s, researchers then showed that ensembles of CILP neural networks, when properly set up, can compute the *modal* fixed-point operator of modal and temporal logic programs. In addition to these results, such ensembles of neural networks were shown to represent the possible world semantics of modal propositional logic, fragments of first order logic and of linear temporal logics. In order to illustrate the computational power of *Connectionist Modal Logics (CML)* and *Connectionist Temporal Logics of Knowledge (CTLK)* [8, 9], researchers were able to learn full solutions to several problems in distributed, multiagent learning and reasoning, including the Muddy Children Puzzle [8] and the Dining Philosophers Problem [26].

By combining temporal logic with modalities, one can represent knowledge and learning evolution in time. This is a key insight, allowing for temporal evolution of both learning and reasoning in time (see Fig. 1). The Figure represents the integrated learning and reasoning process of CTLK. At each time point (or one state of affairs), e.g. t_2, knowledge which the agents are endowed with and what the agents have learned at the previous time t_1 is represented. As time progresses, linear evolution of the agents' knowledge is represented in time as more knowledge about the world (what has been learned) is represented. Fig. 1 illustrates this dynamic property of CTLK, which allows not only the analysis of the current state of affairs but also of how knowledge and learning evolve over time.

Modal and temporal reasoning, when integrated with connectionist learning provide neural-symbolic systems with richer knowledge representation languages and better interpretability. As can be seen in Fig. 1, they enable the construction of more modular deep networks. As argued by Valiant, the construction of cognitive models integrating rich logic-based knowledge representation languages, with robust learning algorithms provide an effective alternative to the construction of semantically sound cognitive neural computational models. It is also argued that a language for describing the algorithms of deep neural networks is needed. Non-classical logics such as logic programming in the context of neuro-symbolic systems, and functional languages used in the context of probabilistic programming are two prominent candidates. In the coming sections, we explain how neural-symbolic systems can be constructed from simple definitions which underline the streamlined integration of knowledge representation, learning, and reasoning in a unified model.

3 Knowledge Representation in Neural Networks

Knowledge representation is the cornerstone of a neural-symbolic system that provides a mapping mechanism between symbolism and connectionism, where logical

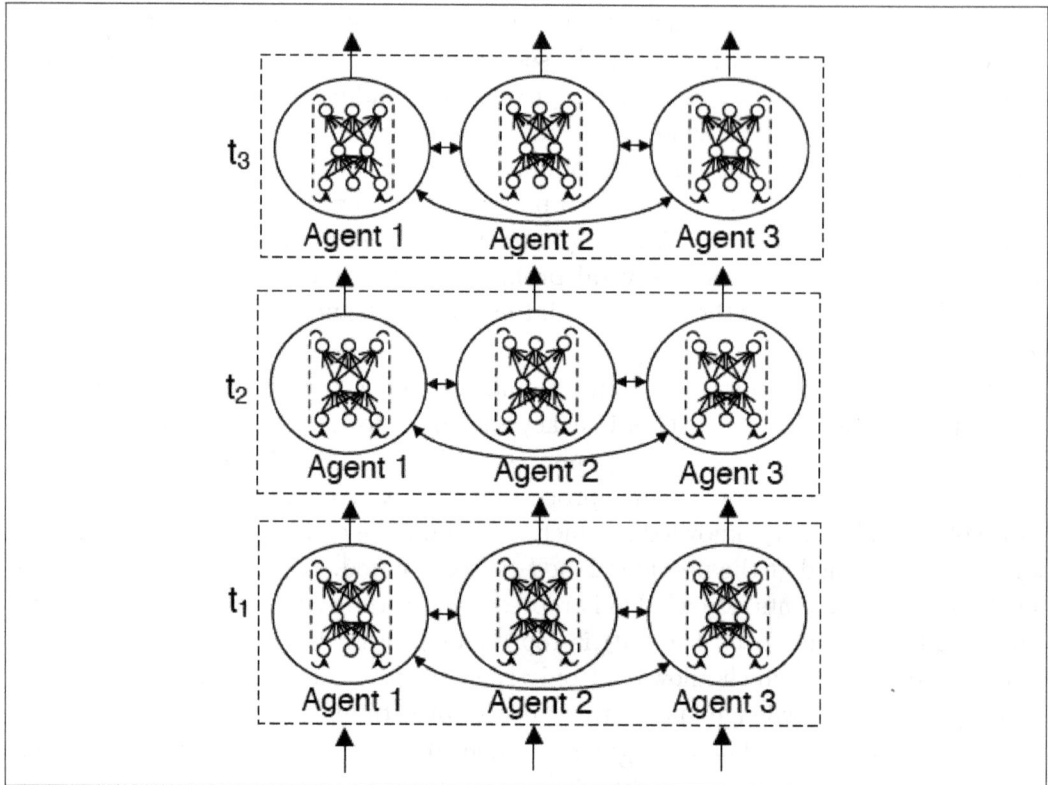

Figure 1: Evolution of Reasoning and Learning in Time

calculus can be carried out exactly or approximately by a neural network. This way, given a trained neural network, symbolic knowledge can be extracted for explaining and reasoning purposes. The representation approaches can be categorised into three main groups: rule-based, formula-based and embedding, which are discussed as follows.

3.1 Propositional Logic

3.1.1 Rule-based Representation

Early work on representation of symbolic knowledge in connectionist networks focused on tailoring the models' parameters to establish an equivalence between input-output mapping function of artificial neural networks (ANN) and logical inference rules. It has been shown that by constraining the weights of a neural network, inference with feedforward propagation can exactly imitate the behaviour of modus

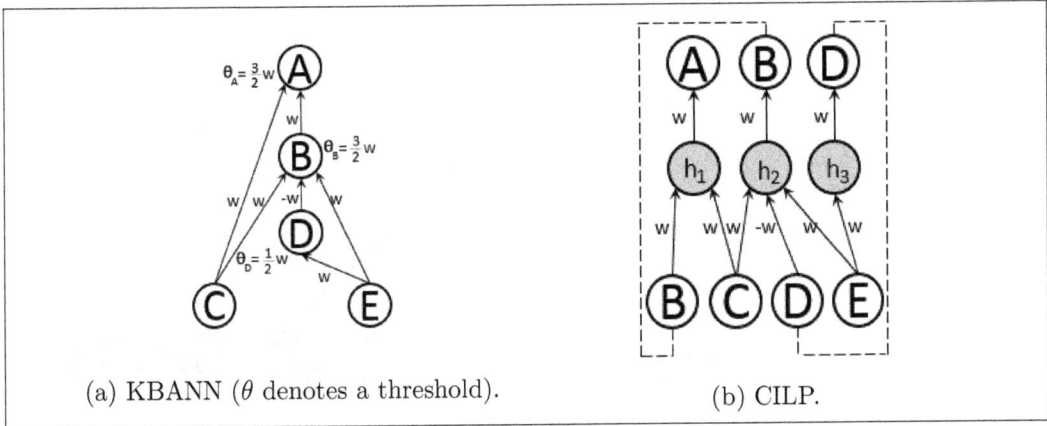

(a) KBANN (θ denotes a threshold).

(b) CILP.

Figure 2: Knowledge representation of $\phi = \{A \leftarrow B \wedge C, B \leftarrow C \wedge \neg D \wedge E, D \leftarrow E\}$ using KBANN and CILP.

ponens [49, 7]. KBANN [49] employs stack of perceptrons to represent the inference rule of logical implications. For example, given a set of rules:

$$\phi = \{A \leftarrow B \wedge C, B \leftarrow C \wedge \neg D \wedge E, D \leftarrow E\} \qquad (1)$$

an ANN can be constructed as in Figure 2a. CILP then generalises the idea by using recurrent networks and bounded continuous units [7]. This representation method allows the use of various data types and more complex sets of rules. With CILP, knowledge given in Eq. (1) can be encoded in a neural network as shown in Figure 2b. In order to adapt this system to first-order logic, CILP++ [15] makes use of techniques from Inductive Logic Programming (ILP). In CILP++, examples and background knowledge are converted into propositional clauses by a bottom-clause propositionalisation technique, which are then encoded into an ANN with recurrent connections as done by CILP.

3.1.2 Formula-based Representation

One issue with KBANN-style rule-based representations is that the discriminative structure of ANNs will only allow a subset of the variables (the consequent of the if-then formula) to be inferred, unless recurrent networks are deployed, with the other variables (the antecedents) being seen as inputs only. This would not represent the behaviour of logical formulas and does not support general reasoning where any variable can be inferred. In order to solve this issue, generative neural networks can be employed as they can treat all variables as non-discriminative. In this formula-based approach, typically associated with restricted Boltzmann machines

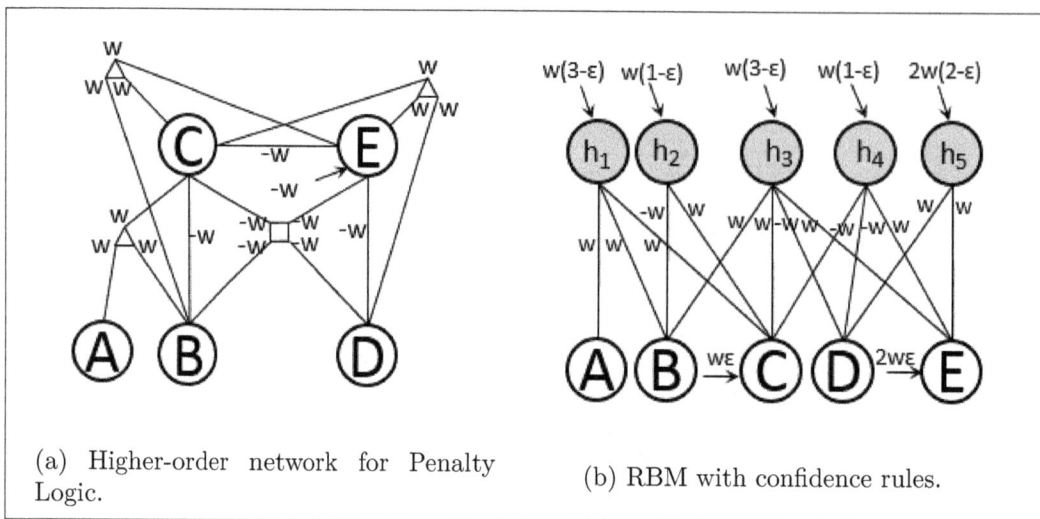

(a) Higher-order network for Penalty Logic.

(b) RBM with confidence rules.

Figure 3: Knowledge representation of $\phi = \{w : A \leftarrow B \wedge C, w : B \leftarrow C \wedge \neg D \wedge E, w : D \leftarrow E\}$ using Penalty logic and Confidence rules

(RBMs) as a building block, the focus is on mapping logical formulas to symmetric connectionist networks, each characterised by an energy function. Early work such as penalty logic [35] proposes a mechanism to represent weighted formulas in energy-based connectionist (Hopfield) networks where maximising satisfiability is equivalent to minimising energy function. Suppose that each formula in the knowledge base (1) is assigned a weight w. Penalty logic constructs a higher-order Hopfield network as shown in Figure 3a. However, inference with such type of network is difficult, while converting the higher-order energy function to a quadratic form is possible but computationally expensive. Recent work on confidence rules [51] proposes an efficient method to represent propositional formulas in restricted Boltzmann machines and deep belief networks where inference and learning become easier. Figure 3b shows an RBM for the knowledge base (1). Nevertheless, learning and reasoning with restricted Boltzmann machines are still complex, making it more difficult to apply formula-based representations than rule-based representations in practice. The main issue has to do with the partition functions of symmetric connectionist networks which cannot be computed analytically. This intractability problem, fortunately, can be ameliorated using sum-product approach as has been shown in [38]. However, it is not yet clear how to apply this idea to RBMs.

3.2 First-order Logic

3.2.1 Propositionalisation

Representation of knowledge in first-order logic in neural networks has been an ongoing challenge, but it can benefit from studies of propositional logic representation 3.1 using propositionalisation techniques [30]. Such techniques allow a first-order knowledge base to be converted into a propositional knowledge base so as to preserve entailment. In neural-symbolic computing, bottom clause prositionalisation (BCP) is a popular approach because bottom clause literals can be encoded directly into neural networks as data features while at the same time presenting semantic meaning.

Early work from [11] employs prositionalisation and feedforward neural networks to learn a clause evaluation function which helps improve the efficiency in exploring large hypothesis spaces. In this approach, the neural network does not work as a standalone ILP system, instead it is used to approximate clause evaluation scores to decide the direction of the hypothesis search. In [36], prositionalisation is used for learning first- order logic in Bayesian networks. Inspired by this work, in [15], the CILP++ system is proposed by integrating bottom clauses and rule-based approach CILP [17], referred to in Section 3.1.1.

The main advantage of propositionalisation is that it is efficient and it fits neural networks well. Also, it does not require first-order formulas to be provided as bottom clauses. However, propositionalisation has serious disadvantages. First, with function symbols, there are infinitely many ground terms. Second, propositionalization seems to generate lots of irrelevant clauses.

3.2.2 Tensorisation

Tensorisation is a class of approaches that embeds first-order logic symbols such as constants, facts and rules into real-valued tensors. Normally, constants are represented as one-hot vectors (first order tensor). Predicates and functions are matrices (second-order tensor) or higher-order tensors.

In early work, embedding techniques were proposed to transform symbolic representations into vector spaces where reasoning can be done through matrix computation [4, 47, 48, 42, 41, 6, 14, 57, 13, 39]. Training embedding systems can be carried out as distance learning using backpropagation. Most research in this direction focuses on representing relational predicates in a neural network. This is known as "relational embedding" [4, 41, 47, 48]. For representation of more complex logical structures, i.e. first order-logic formulas, a system named *Logic Tensor Network* (LTN) [42] is proposed by extending *Neural Tensor Networks* (NTN) [47], a state-

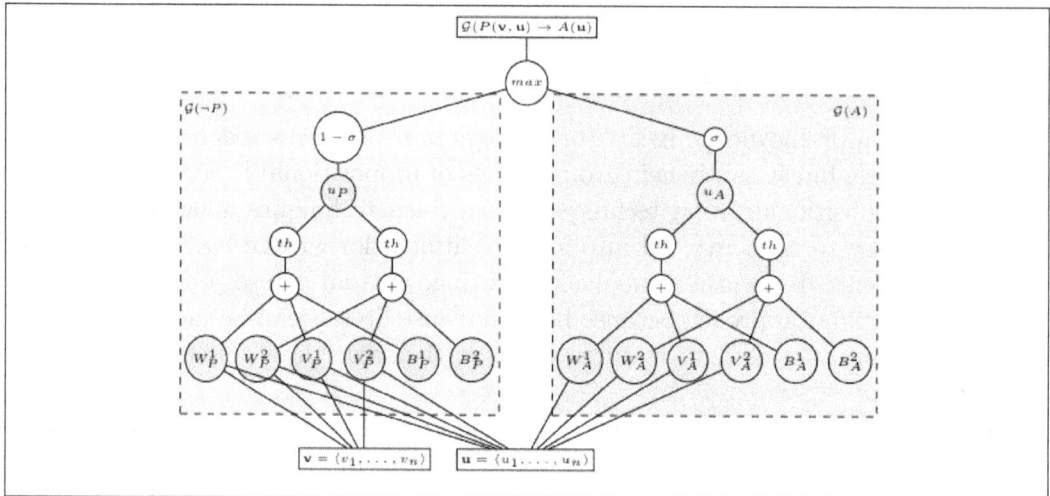

Figure 4: Logic tensor network for $P(x, y) \rightarrow A(y)$ with $\mathcal{G}(x) = \mathbf{v}$ and $\mathcal{G}(y) = \mathbf{u}$; $\mathcal{G}$ are grounding (vector representation) for symbols in first-order language; and the tensor order in this example is 2 [42].

of-the-art relational embedding method. Figure 4 shows an example of LTN for $P(x, y) \rightarrow A(y)$. Related ideas are discussed formally in the context of constraint-based learning and reasoning [19]. Recent research in first-order logic programs has successfully exploited the advantages of distributed representations of logic symbols for efficient reasoning [6], inductive programming [14, 57, 13], and differentiable theorem proving [39].

3.3 Temporal Logic

One of the earliest works on temporal logic and neural networks is CTLK, where ensembles of recurrent neural networks are set up to represent the possible world semantics of linear temporal logics [8]. With single hidden layers and semi-linear neurons, the networks can compute a fixed-point semantics of temporal logic rules. Another work on representation of temporal knowledge is proposed in *Sequential Connectionist Temporal Logic* (SCTL) [5] where CILP is extended to work with the nonlinear auto-regressive exogenous NARX network model. *Neural-Symbolic Cognitive Agents* (NSCA) represent temporal knowledge in recurrent temporal RBMs [34]. Here, the temporal logic rules are modelled in the form of recursive conjunctions represented by recurrent structures of RBMs. Temporal relational knowledge embedding has been studied recently in *Tensor Product Recurrent Neural Network* (TPRN) with applications to question-answering [32].

4 Neural-Symbolic Learning

4.1 Inductive Logic Programming

Inductive logic programming (ILP) can take advantage of the learning capability of neural-symbolic computing to automatically construct a logic program from examples. Normally, approaches in ILP are categorised into *bottom-up* and *top-down* which inspire the development of neural-symbolic approaches accordingly for learning logical rules.

Bottom-up approaches construct logic programs by extracting specific clauses from examples. After that, generalisation procedures are usually applied to search for more general clauses. This is well suited to the idea of propositionalisation discussed earlier in Section 3.2.1. For example, CILP++ [15] employed a bottom clause propositionalisation technique to construct CILP++. In [52], a system called CRILP is proposed by integrating bottom clauses generated from [15] with RBMs. However, both CILP++ and CRILP learn and fine-tune formulas at a propositional level where propositionalisation would generate a large number of long clauses resulting in very large networks. This leaves an open research question of generalising bottom clauses within neural networks that scale well and can extrapolate.

Top-down approaches, on the other hand, construct logic programs from the most general clauses and extend them to be more specific. In neural-symbolic terms, the most popular idea is to take advantage of neural networks' learning and inference capabilities to fine-tune and test the quality of rules. This can be done by replacing logical operations by differentiable operations. For example, in Neural Logic Programming (NLP) [57], learning of rules are based on the differentiable inference of TensorLog [6]. Here, matrix computations are used to soften logic operators where the confidence of conjunctions and confidence of disjunctions are computed as product and sum, respectively. NLP generate rules from facts, starting with the most general ones. In Differentiable Inductive Logic Programming (∂ILP) [14], rules are generated from templates, which are assigned to parameters (weights) to make the loss function between actual conclusions and predicted conclusions from forward chaining differentiable. In [39], Neural Theorem Prover (NTP) is proposed by extending the backward chaining method to be differentiable. It shows that latent predicates from rule templates can be learned through optimisation of their distributed representations. Different from [57, 14, 39] where clauses are generated and then softened by neural networks, in Neural Logic Machines (NLM) [13] the relation of predicates is learned by a neural network where input tensors represent facts (predicates of different arities) from a knowledge base and output tensors represent new facts.

4.2   Horizontal Hybrid Learning

Effective techniques such as deep learning usually require large amounts of data to exhibit statistical regularities. However, in many cases where collecting data is difficult a small dataset would make complex models more prone to overfitting. When prior knowledge is provided, e.g. from domain experts, a neural-symbolic system can offer the advantage of generality by combining logical rules/formulas with data during learning, while at the same time using the data to fine-tune the knowledge. It has been shown that encoding knowledge into a neural network can result in performance improvements [7, 12, 49, 52]. Also, it is evident that using symbolic knowledge can help improve the efficiency of neural network learning [7, 15]. Such effectiveness and efficiency are obtained by encoding logical knowledge as controlled parameters during the training of a model. This technique, in general terms, has been known as learning with logical constraints [19]. Besides, in the case of lacking prior knowledge one can apply the idea of neural-symbolic integration for knowledge transfer learning [51]. The idea is to extract symbolic knowledge from a related domain and transfer it to improve the learning in another domain, starting from a network that does not necessarily have to be instilled with background knowledge. Self-transfer with symbolic-knowledge distillation [23] is also useful as it can enhance several types of deep networks such as convolutional neural networks and recurrent neural networks. Here, symbolic knowledge is extracted from a trained network called "teacher" which then would be used to encoded as regularizers to train a "student" network in the same domain.

4.3   Vertical Hybrid Learning

Studies in neuroscience show that some areas in the brain are used for processing input signals e.g. visual cortices for images [20, 37], while other areas are responsible for logical thinking and reasoning [43]. Deep neural networks can learn high level abstractions from complex input data such as images, audio, and text, which should be useful at making decisions. However, despite that optimisation process during learning being mathematically justified, it is difficult for humans to comprehend how a decision has been made during inference time. Therefore, placing a logic network on top of a deep neural network to learn the relations of those abstractions, can help the system to be able to explain itself. In [12], a Fast-RCNN [18] is used for bounding-box detection of parts of objects and on top of that, a Logic Tensor Network is used to reason about relations between parts of objects and types of such objects. In such work, the perception part (Fast-RCNN) is fixed and learning is carried out in the reasoning part (LTN). In a related approach, called DeepProbLog, end-to-end learning and reasoning have been studied [28] where outputs of neural networks are used as "neural predicates" for ProbLog [10].

5 Neural-symbolic Reasoning

Reasoning is an important feature of a neural-symbolic system and has recently attracted much attention from the research community [14]. Various attempts have been made to perform reasoning within neural networks, both model-based and theorem proving approaches. In neural-symbolic integration the main focus is the integration of reasoning and learning, so that a model-based approach is preferred. Most theorem proving systems based on neural networks, including first-order logic reasoning systems such as SHRUTI [56], have been unable to perform learning as effectively as end-to-end differentiable learning systems. On the other hand, model-based approaches have been shown implementable in neural networks in the case of nonmonotonic, intuitionistic and propositional modal logic, as well as abductive reasoning and other forms of human reasoning [2, 5]. As a result, the focus of neural-symbolic computation has changed from performing symbolic reasoning in neural networks, such as for example implementing the logical unification algorithm in a neural network, to the combination of learning and reasoning, in some cases with a much more loosely-defined approach rather than full integration, whereby a hybrid system will contain different components which may be neural or symbolic and which communicate with each other.

5.1 Forward and Backward chaining

Forward chaining and backward chaining are two popular inference techniques for logic programs and other logical systems. In the case of neural-symbolic systems forward and backward chainings are both in general implemented by feedforward inference.

Forward chaining generates new facts from the head literals of the rules using existing facts in the knowledge base. For example, in [34], a "Neural-symbolic Cognitive Agent" shows that it is possible to perform online learning and reasoning in real-world scenarios, where temporal knowledge can be extracted to reason about driving skills [34]. This can be seen as forward chaining over time. In ∂ILP [14], a differentiable function is defined for each clause to carry out a single step of forward chaining. Similar to this, NLM [13] employs neural networks as a differentiable chain for forward inference. Different from ∂ILP, NLM represent the outputs and inputs of neural networks as grounding tensors of predicates for existing facts and new facts respectively.

Backward chaining, on the other hand, searches backward from a goal in the knowledge base to determine whether a query is derivable or not. This form a tree search starts from the query and expands further to the literals in the body of the rules whose heads match the query. TensorLog [6] implements backward chaining using neural networks as symbols. The idea is based on stochastic logic programs [31], and soft logic is applied to transform the hypothesis search into a chain of matrix operations. In NTP, a neural system is constructed recursively for backward chaining and unification where AND and OR operators are represented as networks. In general, backward (goal-directed) reasoning is considerably harder to achieve in neural networks than forward reasoning. This is another current line of research within neuro-symbolic computation and AI.

5.2 Approximate Satisfiability

Inference in the case of logic programs with arbitrary formulas is more complex. In general, one may want to search over the hypothesis space to find a solution that satisfies (mostly) the formulas and facts in the knowledge base. Exact inference, that is, reasoning maximising satisfiability, is NP-hard. For this reason, some neural-symbolic systems offer a mechanism of approximate satisfiability. Tensor logic networks are trained to approximate the best satisfiability [42] making inference efficient with feedforward propagation. This has made LTNs applicable successfully to the Pascal data set and image understanding [12]. Penalty logic shows an equivalence between minimising violation and minimising energy functions of symmetric connectionist networks [35]. Confidence rules, another approximation approach, shows the relation between sampling in restricted Boltzmann machines and search for truth-assignments which maximise satisfiability. The use of confidence rules also allows one to measure how confident a neural network is in its own answers. Based on that, neural-symbolic system "confidence rule inductive logic programming (CRILP)" was constructed and applied to inductive logic programming [52].

5.3 Relationship reasoning

Relational embedding systems have been used for reasoning about relationships between entities. Technically, this has been done by searching for the answer to a query that gives the highest grounding score [4, 3, 47, 48]. Deep neural networks are also employed for visual reasoning where they learn and infer relationships and features of multiple objects in images [41, 58, 29].

6   Neural-symbolic Explainability

The (re)emergence of deep networks has again raised the question of explainability. The complex structure of a deep neural network turns them into a powerful learning system if one can correctly engineer its components such as type of hidden units, regularisation and optimization methods. However, limitations of some AI applications have heightened the need for explainability and interpretability of deep neural networks. More importantly, besides improving deep neural networks for better applications one should also look for the benefits that deep networks can offer in terms of knowledge acquisition.

6.1   Knowledge Extraction

Explainability is a promising capability of neural-symbolic systems where the behaviour of a connectionist network can be represented in a set of human-readable expressions. In early work, the demand for solving "black-box" issues of neural networks has motivated a number of rules extraction methods. Most of them are discussed in the surveys [1, 24, 55]. These attempts were to search for logic rules from a trained network based on four criteria: (a) accuracy, (b) fidelity, (c) consistency and (d) comprehensibility [1]. In [17], a sound extraction approach based on partially ordered sets is proposed to narrow the search of logic rules. However, such combinatorial approaches do not scale well to deal with the dimensionality of current networks. As a result, gradually less attention has been paid to knowledge extraction until recently when the combination of global and local approaches started to be investigated. The idea here is either to create modular networks with rule extraction applying to specific modules or to consider rule extraction from specific layers only.

In [50, 51], it has been shown that while extracting conjunctive clauses from the first layer of a deep belief network is fast and effective, extraction in higher layers results in a loss of accuracy. A trained deep network can be employed instead for extraction of soft-logic rules which is less formal but more flexible [23]. Extraction of temporal rules have been studied in [34] and generated semantic relations of domain variables over time. Besides formal logical knowledge, hierarchical Boolean expressions can be learned from images for object detection and recognition [44].

6.2   Natural Language Generation

For explainability purposes, another approach couples a deep network with sequence models to extract natural language knowledge [22]. In [4], instead of investigating the parameters of a trained model, relational knowledge extraction is proposed where predicates are obtained by performing inference of a trained embedding network on text data.

6.3 Program Synthesis

In the field of Program Induction, neuro-symbolic program synthesis (NSPS) has been proposed to construct computer programs on an incremental fashion using a large amount of input-output samples [33]. A neural network is employed to represent partial trees in a domain-specific language are tree nodes, symbols and rules are vector representations. Explainability can be achieved through the tree-based structure of the network. Again, this shows that the integration of neural networks and symbolic representation is indeed a solution for both scalability and explainability.

7 Conclusions

In this paper, we highlighted the key ideas and principles of neural-symbolic computing. In order to do so, we illustrated the main methodological approaches which allow for the integration of effective neural learning with sound symbolic-based, knowledge representation and reasoning methods. One of the principles we highlighted in the paper is the sound mapping between symbolic rules and neural networks provided by neural-symbolic computing methods. This mapping allows several knowledge representation formalisms to be used as background knowledge for potentially large-scale learning and efficient reasoning. This interplay between efficient neural learning and symbolic reasoning opens relevant possibilities towards richer intelligent systems. The comprehensibility and compositionality of neural-symbolic systems, offered by building networks with a logical structure, allows for integrated learning and reasoning under different logical systems. This opens several interesting research lines, in which learning is endowed with the sound semantics of diverse logics. This, in turn, contributes towards the development of explainable and accountable AI and machine learning-based systems and tools.

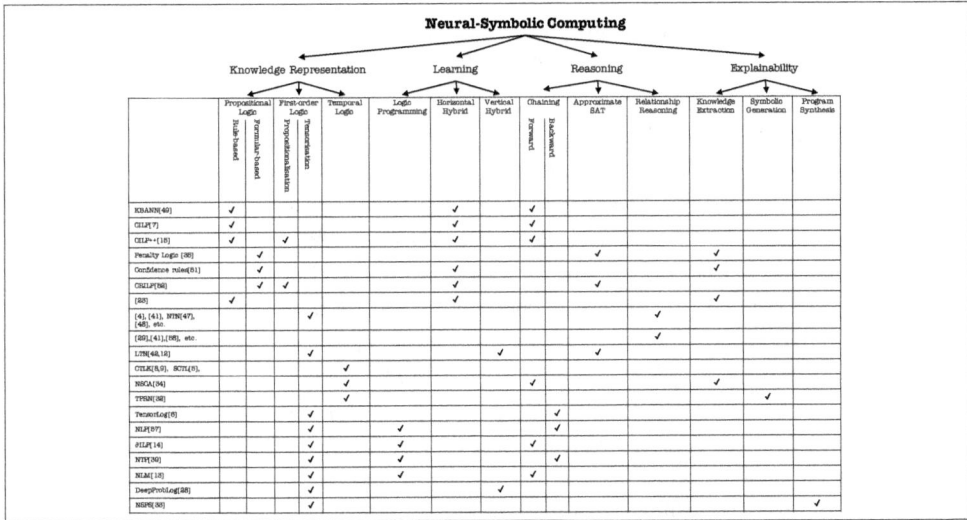

Figure 5: Summary of Neural-symbolic Techniques

	Propositional Logic			First-order Logic	Temporal Logic	Logic Programming	Horizontal Hybrid	Vertical Hybrid	Chaining		Approximate SAT	Relationship Reasoning	Knowledge Extraction	Symbolic Generation	Program Synthesis
	Rule-based	Formulae-based	Propositionalization	Tensorisation					Forward	Backward					
KBANN[49]	✓						✓		✓						
CILP[7]	✓						✓		✓						
CILP++[18]	✓		✓				✓		✓						
Penalty Logic [39]		✓									✓		✓		
Confidence rules[51]		✓					✓						✓		
CIL2P[56]		✓	✓				✓						✓		
[20]	✓						✓						✓		
[4], [41], NTN[47], [48], etc.				✓								✓			
[29],[41],[98], etc.												✓			
LTN[48,13]				✓					✓		✓				
CTLK[8,9], SCTL[6],					✓										
NSCA[34]					✓				✓				✓		
TPRN[32]					✓									✓	
TensorLog[5]				✓						✓					
NLP[97]				✓		✓				✓					
∂ILP[14]				✓		✓			✓						
NTP[50]				✓		✓				✓					
NLM[13]				✓		✓			✓						
DeepProbLog[38]				✓				✓							
NGS[88]				✓											✓

References

[1] R. Andrews, J. Diederich, and A. Tickle. Survey and critique of techniques for extracting rules from trained artificial neural networks. *Know.-Based Syst.*, 8(6):373–389, December 1995.

[2] S. Bader and P. Hitzler. Dimensions of neural-symbolic integration: a structured survey. In S. Artemov, H. Barringer, A. d'Avila Garcez, L. Lamb, and J. Woods, editors, *We Will Show Them! Essays in Honour of Dov Gabbay*, 2005.

[3] A. Bordes, X. Glorot, J. Weston, and Y. Bengio. Joint learning of words and meaning representations for open-text semantic parsing. In *AISTATS*, pages 127–135, 2012.

[4] A. Bordes, J. Weston, R. Collobert, and Y. Bengio. Learning structured embeddings of knowledge bases. In *AAAI'11*, 2011.

[5] R. Borges, A. d'Avila Garcez, and L.C. Lamb. Learning and representing temporal knowledge in recurrent networks. *IEEE Trans. Neural Networks*, 22(12):2409–2421, 2011.

[6] William W. Cohen, Fan Yang, and Kathryn Mazaitis. Tensorlog: Deep learning meets probabilistic dbs. *CoRR*, abs/1707.05390, 2017.

[7] A. d'Avila Garcez and G. Zaverucha. The connectionist inductive learning and logic programming system. *Appl. Intelligence*, 11(1):5977, 1999.

[8] A.S. d'Avila Garcez and L.C. Lamb. Reasoning about time and knowledge in neural symbolic learning systems. In *NIPS*, pages 921–928, 2003.

[9] A.S. d'Avila Garcez and L.C. Lamb. A connectionist computational model for epistemic and temporal reasoning. *Neur. Computation*, 18(7):1711–1738, 2006.

[10] Luc De Raedt, Angelika Kimmig, and Hannu Toivonen. Problog: A probabilistic prolog and its application in link discovery. In *Proceedings of the 20th International Joint Conference on Artifical Intelligence*, IJCAI'07, pages 2468–2473, San Francisco, CA, USA, 2007. Morgan Kaufmann Publishers Inc.

[11] Frank DiMaio and Jude Shavlik. Learning an approximation to inductive logic programming clause evaluation. In Rui Camacho, Ross King, and Ashwin Srinivasan, editors, *Inductive Logic Programming*, pages 80–97, Berlin, Heidelberg, 2004. Springer Berlin Heidelberg.

[12] I. Donadello, L. Serafini, and A. S. d'Avila Garcez. Logic tensor networks for semantic image interpretation. In *IJCAI-17*, pages 1596–1602, 2017.

[13] Honghua Dong, Jiayuan Mao, Tian Lin, Chong Wang, Lihong Li, and Denny Zhou. Neural logic machines. 2019.

[14] R. Evans and E. Grefenstette. Learning explanatory rules from noisy data. *JAIR*, 61:1–64, 2018.

[15] M. França, G. Zaverucha, and A. Garcez. Fast relational learning using bottom clause propositionalization with artificial neural networks. *Mach. Learning*, 94(1):81–104, 2014.

[16] A. d'Avila Garcez, L.C. Lamb, and D.M. Gabbay. *Neural-Symbolic Cognitive Reasoning*.

Springer, 2009.

[17] A.S. d'Avila Garcez, D. Gabbay, and K.. Broda. *Neural-Symbolic Learning System: Foundations and Applications.* Springer, 2002.

[18] Ross Girshick. Fast r-cnn. In *Proceedings of the 2015 IEEE International Conference on Computer Vision (ICCV)*, ICCV '15, pages 1440–1448, Washington, DC, USA, 2015. IEEE Computer Society.

[19] Marco Gori. *Machine Learning: A Constraint-Based Approach.* Morgan Kaufmann, 2018.

[20] Kalanit Grill-Spector and Rafael Malach. The human visual cortex. *Annual Review of Neuroscience*, 27(1):649–677, 2004.

[21] B. Hammer and P. Hitzler, editors. *Perspectives of Neural-Symbolic Integration.* Springer, 2007.

[22] Lisa Anne Hendricks, Zeynep Akata, Marcus Rohrbach, Jeff Donahue, Bernt Schiele, and Trevor Darrell. Generating visual explanations. In *Computer Vision - ECCV 2016 - 14th European Conference*, pages 3–19, 2016.

[23] Z. Hu, X. Ma, Z. Liu, E. Hovy, and E. Xing. Harnessing deep neural networks with logic rules. In *ACL*, 2016.

[24] Henrik Jacobsson. Rule extraction from recurrent neural networks: A taxonomy and review. *Neural Comput.*, 17(6):1223–1263, June 2005.

[25] R. Khardon and D. Roth. Learning to reason. *J. ACM*, 44(5), 1997.

[26] L.C. Lamb, R.V. Borges, and A.S. d'Avila Garcez. A connectionist cognitive model for temporal synchronisation and learning. In *AAAI*, pages 827–832, 2007.

[27] Y. LeCun, Y. Bengio, and G. Hinton. Deep learning. *Nature*, 521(7553):436–444, 2015.

[28] Robin Manhaeve, Sebastijan Dumancic, Angelika Kimmig, Thomas Demeester, and Luc De Raedt. Deepproblog: Neural probabilistic logic programming. In S. Bengio, H. Wallach, H. Larochelle, K. Grauman, N. Cesa-Bianchi, and R. Garnett, editors, *Advances in Neural Information Processing Systems 31*, pages 3749–3759. Curran Associates, Inc., 2018.

[29] Jiayuan Mao, Chuang Gan, Pushmeet Kohli, Joshua B. Tenenbaum, and Jiajun Wu. The Neuro-Symbolic Concept Learner: Interpreting Scenes, Words, and Sentences From Natural Supervision. In *International Conference on Learning Representations*, 2019.

[30] Stephen Muggleton. Inverse entailment and progol. *New Generation Computing*, 13(3):245–286, Dec 1995.

[31] Stephen Muggleton. Stochastic logic programs. In *New Generation Computing*. Academic Press, 1996.

[32] H. Palangi, P. Smolensky, X. He, and L. Deng. Question-answering with grammatically-interpretable representations. In *AAAI*, 2018.

[33] E. Parisotto, A.-R. Mohamed, R. Singh, L. Li, D. Zhou, and P. Kohli. Neuro-symbolic program synthesis. In *ICLR*, 2017.

[34] L. de Penning, A. d'Avila Garcez, L.C. Lamb, and J-J. Meyer. A neural-symbolic cognitive agent for online learning and reasoning. In *IJCAI*, pages 1653–1658, 2011.

[35] G. Pinkas. Reasoning, nonmonotonicity and learning in connectionist networks that capture propositional knowledge. *Artif. Intell.*, 77(2):203–247, 1995.

[36] Cristiano Grijó Pitangui and Gerson Zaverucha. Learning theories using estimation distribution algorithms and (reduced) bottom clauses. In Stephen H. Muggleton, Alireza Tamaddoni-Nezhad, and Francesca A. Lisi, editors, *Inductive Logic Programming*, pages 286–301, Berlin, Heidelberg, 2012. Springer Berlin Heidelberg.

[37] Tomaso A. Poggio and Fabio Anselmi. *Visual Cortex and Deep Networks: Learning Invariant Representations*. The MIT Press, 1st edition, 2016.

[38] H. Poon and P. Domingos. Sum-product networks: A new deep architecture. In *2011 IEEE Int. Conf. on Computer Vision Workshops (ICCV Workshops)*, 2011.

[39] Tim Rocktäschel and Sebastian Riedel. Learning knowledge base inference with neural theorem provers. In *Proceedings of the 5th Workshop on Automated Knowledge Base Construction*, pages 45–50, San Diego, CA, June 2016. Association for Computational Linguistics.

[40] S.J. Russell, S. Hauert, R. Altman, and M. Veloso. Ethics of artificial intelligence: Four leading researchers share their concerns and solutions for reducing societal risks from intelligent machines. *Nature*, 521:415–418, 2015.

[41] A. Santoro, D. Raposo, D. Barrett, M. Malinowski, R. Pascanu, P. Battaglia, and T. Lillicrap. A simple neural network module for relational reasoning. In *NIPS*, 2017.

[42] Luciano Serafini and Artur S. d'Avila Garcez. Learning and reasoning with logic tensor networks. In *AI*IA*, pages 334–348, 2016.

[43] Ehsan Shokri-Kojori, Michael A. Motes, Bart Rypma, and Daniel C. Krawczyk. The network architecture of cortical processing in visuo-spatial reasoning. *Scientific Reports*, 2, 2012.

[44] Z. Si and S. C. Zhu. Learning and-or templates for object recognition and detection. *IEEE Trans. Pattern Analysis and Mach. Intell.*, 35(9):2189–2205, Sept 2013.

[45] D. Silver, J. Schrittwieser, K. Simonyan, I. Antonoglou, A. Huang, A. Guez, T. Hubert, L. Baker, M. Lai, A. Bolton, Y. Chen, T. Lillicrap, F. Hui, L. Sifre, G. van den Driessche, T. Graepel, and D. Hassabis. Mastering the game of go without human knowledge. *Nature*, 550(354), 2017.

[46] P. Smolensky. Constituent structure and explanation in an integrated connectionist/symbolic cognitive architecture. In *Connectionism: Debates on Psychological Explanation*. 1995.

[47] R. Socher, D. Chen, C. Manning, and A. Ng. Reasoning with neural tensor networks for knowledge base completion. In *NIPS*, pages 926–934. 2013.

[48] I. Sutskever and G. Hinton. Using matrices to model symbolic relationship. In *NIPS*. 2009.

[49] G. Towell and J. Shavlik. Knowledge-based artificial neural networks. *Artif. Intel.*, 70:119–165, 1994.

[50] S. Tran and A. d'Avila Garcez. Knowledge extraction from deep belief networks for images. In *IJCAI-2013 Workshop on Neural-Symbolic Learning and Reasoning*, 2013.

NEURAL-SYMBOLIC COMPUTING

[51] S. Tran and A. Garcez. Deep logic networks: Inserting and extracting knowledge from deep belief networks. *IEEE T. Neur. Net. Learning Syst.*, (29):246–258, 2018.

[52] Son N. Tran. Propositional knowledge representation and reasoning in restricted boltzmann machines. *CoRR*, abs/1705.10899, 2018.

[53] L. Valiant. Knowledge infusion. In *AAAI*, 2006.

[54] L.G. Valiant. Three problems in computer science. *J. ACM*, 50(1):96–99, 2003.

[55] Qinglong Wang, Kaixuan Zhang, Alexander G. Ororbia II, Xinyu Xing, Xue Liu, and C. Lee Giles. An empirical evaluation of rule extraction from recurrent neural networks. *Neural Computation*, 30(9):2568–2591, 2018.

[56] Carter Wendelken and Lokendra Shastri. Multiple instantiation and rule mediation in SHRUTI. *Connect. Sci.*, 16(3):211–217, 2004.

[57] Fan Yang, Zhilin Yang, and William W Cohen. Differentiable learning of logical rules for knowledge base reasoning. In I. Guyon, U. V. Luxburg, S. Bengio, H. Wallach, R. Fergus, S. Vishwanathan, and R. Garnett, editors, *Advances in Neural Information Processing Systems 30*, pages 2319–2328. Curran Associates, Inc., 2017.

[58] Kexin Yi, Jiajun Wu, Chuang Gan, Antonio Torralba, Pushmeet Kohli, and Josh Tenenbaum. Neural-symbolic vqa: Disentangling reasoning from vision and language understanding. In S. Bengio, H. Wallach, H. Larochelle, K. Grauman, N. Cesa-Bianchi, and R. Garnett, editors, *Advances in Neural Information Processing Systems 31*, pages 1031–1042. Curran Associates, Inc., 2018.

Received 7 June 2018

Semi-Supervised Learning using Differentiable Reasoning

Emile van Krieken
VU University Amsterdam
`e.van.krieken@vu.nl`

Erman Acar
VU University Amsterdam
`erman.acar@vu.nl`

Frank van Harmelen
VU University Amsterdam
`Frank.van.Harmelen@vu.nl`

Abstract

We introduce Differentiable Reasoning (DR), a novel semi-supervised learning technique which uses relational background knowledge to benefit from unlabeled data. We apply it to the Semantic Image Interpretation (SII) task and show that background knowledge provides significant improvement. We find that there is a strong but interesting imbalance between the contributions of updates from Modus Ponens (MP) and its logical equivalent Modus Tollens (MT) to the learning process, suggesting that our approach is very sensitive to a phenomenon called the Raven Paradox [10]. We propose a solution to overcome this situation.

1 Introduction

Semi-supervised learning is a common class of methods for machine learning tasks where we consider not just labeled data, but also make use of unlabeled data [2]. This can be very beneficial for training in tasks where labeled data is much harder to acquire than unlabeled data.

We thank the anonymous reviewers of NeSy for their valueable feedback. Additionally we thank Luciano Serafini, Jasper Driessens, Fije Overeem and also the other attendees of NeSy for the great discussions.

One such task is *Semantic Image Interpretation (SII)* in which the goal is to generate a semantic description of the objects on an image [7]. This description is represented as a labeled directed graph, which is known as a *scene graph* [13]. An example of a labeled dataset for this problem is VisualGenome [15] which contains 108,077 images to train 156,722 different unary and binary predicates. The binary relations in particular make this dataset very sparse, as there are many different pairs of objects that could be related. However, a far larger, though unfortunately unlabeled, dataset like ImageNet [24] contains over 14 million different pictures. Because it is so much larger, it will have many examples of interactions that are not present in VisualGenome. We show that it is possible to improve the performance of a simple classifier on the SII task significantly by adding the satisfaction of a first-order logic (FOL) knowledge base to the supervised loss function. The computation of this satisfaction uses an unlabeled dataset as its domain.

For this purpose, we introduce a statistical relational learning framework called *Differentiable Reasoning* (DR) in Section 2, as our primary contribution. DR uses simple logical formulas to deduce new training examples in an unlabeled dataset. This is done by adding a differentiable loss term that evaluates the truth value of the formulas.

In the experimental analysis, we find that the gradient updates using the Modus Ponens (MP) and Modus Tollens (MT) rules are disproportionate. That is, MT often strongly dominates MP in the learning process. Such behavior suggests that our approach is highly sensitive to the Raven Paradox [10]. It refers to the phenomenon that the observations obtained from "All ravens are black" are dominated by its logically equivalent "All non-black things are non-ravens". Indeed, this is closely related to the *material implication* which caused a lot of discussion throughout the history of logic and philosophy [8]. Our second main contribution relies on its investigation in Section 2.4, and our proposal to cope with it. Finally, we show results on a simple dataset in Section 3 and analyze the behavior of the Raven Paradox in Section 4. Related works and conclusion closes the paper.

2 Differentiable Reasoning

2.1 Basics and Notation

We assume a knowledge base $\mathcal{K}$ is given in a relational logic language, where a formula $\varphi \in \mathcal{K}$ is built from predicate symbols $\mathsf{P} \in \mathcal{P}$, a finite set $\mathcal{D}$ of objects (also called constants) $o \in \mathbb{R}^m$ with $m \in \mathbb{Z}^+$, and variables $x \in \mathcal{V}$, in the usual way (see [28]). We also assume that every $\varphi \in \mathcal{K}$ is in Skolem normal form. For a vector of objects and variables, we use boldfaced $\mathbf{o}$ and $\mathbf{x}$, respectively. A ground atom is

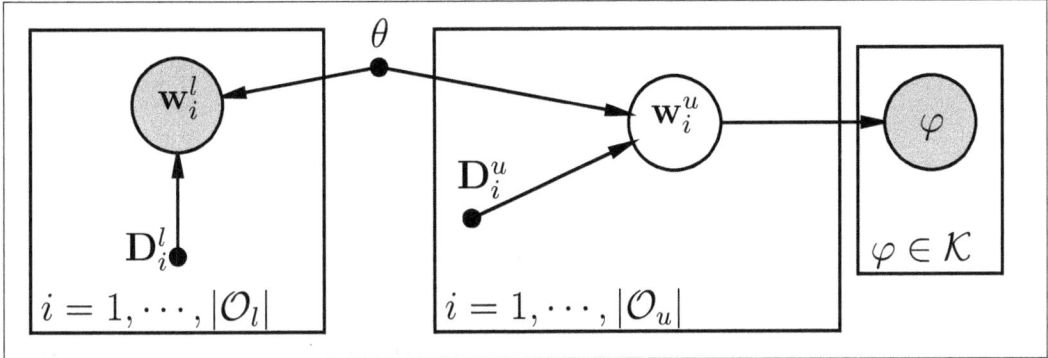

Figure 1: The Bayesian network describing the joint probability $p(\mathcal{W}_l, \mathcal{W}_u, \mathcal{K} | \mathcal{O}_l, \mathcal{O}_u, \boldsymbol{\theta})$. The left plate is the supervised classification likelihood and the right plates the unsupervised part in which we calculate the probability of the formulas $\mathcal{K}$. The parameters $\boldsymbol{\theta}$ are shared in both parts.

a formula with no logical connective and no variables, e.g., partOf($cushion, chair$) where partOf $\in \mathcal{P}$ and $cushion, chair \in \mathcal{D}$. Given a subset $D_i \subseteq \mathcal{D}$, a $Herbrand\ base$ $\mathbf{A}_i$ corresponding to D_i is the set of all ground atoms generated from D_i and $\mathcal{P}$. A $world$ (often called a $Herbrand\ interpretation$) $\mathbf{w}_i$ for D_i assigns a binary truth value to each ground atom $\mathsf{P}(\mathbf{o}) \in \mathbf{A}_i$ i.e., $\mathbf{w}_i(\mathsf{P}(\mathbf{o})) \in \{0, 1\}$.

Each predicate P has a corresponding differentiable function $f_\mathsf{P}^{\boldsymbol{\theta}}(\mathbf{o}) \in \mathbb{R}^{\alpha(\mathsf{P}) \times m} \to [0, 1]$ parameterized by $\boldsymbol{\theta}$ (a vector of reals) with $\alpha(\mathsf{P})$ being the arity of P, which calculates the probability of $\mathsf{P}(\mathbf{o})$. This function could be, for instance, a neural network.

Next, we define a Bernouilli distribution function over worlds as follows

$$p(\mathbf{w}_i | \boldsymbol{\theta}, D_i) = \prod_{\mathsf{P}(\mathbf{o}) \in \mathbf{A}_i} f_\mathsf{P}^{\boldsymbol{\theta}}(\mathbf{o})^{\mathbf{w}_i(\mathsf{P}(\mathbf{o}))} \cdot (1 - f_\mathsf{P}^{\boldsymbol{\theta}}(\mathbf{o}))^{1-\mathbf{w}_i(\mathsf{P}(\mathbf{o}))} \tag{1}$$

where $\mathbf{w}_i(\mathsf{P}(\mathbf{o}))$ (similarly, $1 - \mathbf{w}_i(\mathsf{P}(\mathbf{o}))$) refers to the exponent. Given some world $\mathbf{w_i}$, the $valuation\ function$ $v(\varphi, \mathbf{w_i})$ is 1 if φ is true in that world, that is, $\mathbf{w}_i \models \varphi$, and 0 otherwise.

Next, we explain the domain we use in this article. We have a dataset $\mathcal{D}$ partitioned into two parts: a labeled dataset $\mathcal{D}_l = \langle \mathcal{O}_l, \mathcal{W}_l \rangle$ and an unlabeled dataset $\mathcal{D}_u = \langle \mathcal{O}_u, \emptyset \rangle$ where both $\mathcal{O}_l$ and $\mathcal{O}_u$ are sets of finite domains D_i, and $\mathcal{W}_l$ is a set containing the correct world $\mathbf{w}_i^{l*}$ for all pictures i.

In Figure 1 we illustrate the Bayesian network associated with this problem. The left plate denotes the usual supervised data likelihood $p(\mathcal{W}_l | \mathcal{O}_l, \boldsymbol{\theta})$ and the right plates denote the probabilities of the truth values of the formulas $\varphi \in \mathcal{K}$ using

$p(\mathcal{K}|\mathcal{O}_u, \boldsymbol{\theta})$.

It is important to note that the true worlds $\mathbf{w}_i^{u*}$ of the unlabeled dataset are not known, that is, they are latent variables and they have to be marginalized over. The formulas in knowledge base $\mathcal{K}$ are all assumed to be true. We can now obtain the optimization problem that we can solve using gradient descent as

$$\boldsymbol{\theta}^* = \arg\max_{\boldsymbol{\theta}} p(\mathcal{W}_l|\mathcal{O}_l, \boldsymbol{\theta}) \cdot p(\mathcal{K}|\mathcal{O}_u, \boldsymbol{\theta}) \tag{2}$$

$$= \arg\max_{\boldsymbol{\theta}} \prod_{i=1}^{|\mathcal{O}_l|} p(\mathbf{w}_i^{l*}|D_i^l, \boldsymbol{\theta}) \cdot \prod_{i=1}^{|\mathcal{O}_u|} \sum_{\mathbf{w}_i^u} p(\mathbf{w}_i^u|D_i^u, \boldsymbol{\theta}) \cdot \prod_{\varphi \in \mathcal{K}} v(\varphi, \mathbf{w}_i^u) \tag{3}$$

$$= \arg\min_{\boldsymbol{\theta}} - \sum_{i=1}^{|\mathcal{O}_l|} \log p(\mathbf{w}_i^{l*}|D_i^l, \boldsymbol{\theta})$$

$$- \sum_{i=1}^{|\mathcal{O}_u|} \log \left(\sum_{\mathbf{w}_i^u} p(\mathbf{w}_i^u|D_i^u, \boldsymbol{\theta}) \cdot \prod_{\varphi \in \mathcal{K}} v(\varphi, \mathbf{w}_i^u) \right) \tag{4}$$

where in the last step we take the log and minimize with respect to the negative value. The optimization problem in Equation 4 consists of two terms. The first is the cross-entropy loss for supervised labeled data. The second can be understood as follows: A world entails a (full) knowledge base (i.e., $\mathbf{w} \models \mathcal{K}$) if $\mathbf{w} \models \varphi$ holds for all $\varphi \in \mathcal{K}$ (that is, the product of their valuations is 1). For each domain D_i, we then find the sum of the probabilities of worlds that entail the knowledge base. This is an example of what we call the *differentiable reasoning* loss. The general differentiable reasoning objective is given as

$$\boldsymbol{\theta}^* = \arg\min_{\boldsymbol{\theta}} - \sum_{i=1}^{|\mathcal{O}_l|} \log p(\mathbf{w}_i^{l*}|D_i^l, \boldsymbol{\theta}) + \mathcal{L}_{DR}(\boldsymbol{\theta}; \mathcal{K}, \mathcal{O}_u). \tag{5}$$

2.2 Differentiable Reasoning Using Product Real Logic

The marginalization over all possible worlds $\mathbf{w}_i^u$ requires $2^{|\mathbf{A}_i|}$ combinations, so it is exponential in the size of the Herbrand base. Therefore, the problem of finding the sum of the probabilities $p(\mathbf{w}_i|\boldsymbol{\theta})$ for all worlds $\mathbf{w}_i$ that entail the knowledge base $\mathcal{K}$ is #P-complete [23] Instead, we shall perform a much simpler computation defined over logical formulas and the parameters $\boldsymbol{\theta}$ as follows:

$$\mathcal{L}_{DR}(\boldsymbol{\theta}; \mathcal{K}, \mathcal{O}_u) = \sum_{\varphi \in \mathcal{K}} \mathcal{L}(\boldsymbol{\theta}; \varphi, \mathcal{O}_u) \tag{6}$$

$$\mathcal{L}(\boldsymbol{\theta}; \forall \mathbf{x} \phi, \mathcal{O}_u) = - \sum_{D \in \mathcal{O}_u, \mathbf{o} \in D} \log \hat{p}(\phi | \mathbf{x} = \mathbf{o}, \boldsymbol{\theta}) \tag{7}$$

$$\hat{p}(\mathsf{P}(x_1, ..., x_{\alpha(\mathsf{P})}) | \mathbf{x} = \mathbf{o}, \boldsymbol{\theta}) = f_{\mathsf{P}}^{\boldsymbol{\theta}}(o_1, ..., o_{\alpha(\mathsf{P})}) \tag{8}$$

$$\hat{p}(\neg \phi | \mathbf{x} = \mathbf{o}, \boldsymbol{\theta}) = 1 - \hat{p}(\phi | \mathbf{x} = \mathbf{o}, \boldsymbol{\theta}) \tag{9}$$

$$\hat{p}(\phi \wedge \psi | \mathbf{x} = \mathbf{o}, \boldsymbol{\theta}) = \hat{p}(\phi | \mathbf{x} = \mathbf{o}, \boldsymbol{\theta}) \cdot \hat{p}(\psi | \mathbf{x} = \mathbf{o}, \boldsymbol{\theta}) \tag{10}$$

$$\hat{p}(\phi \vee \psi | \mathbf{x} = \mathbf{o}, \boldsymbol{\theta}) = \hat{p}(\neg(\neg \phi \wedge \neg \psi) | \mathbf{x} = \mathbf{o}, \boldsymbol{\theta}) \tag{11}$$

$$\hat{p}(\phi \to \psi | \mathbf{x} = \mathbf{o}, \boldsymbol{\theta}) = \hat{p}(\neg \phi \vee \psi | \mathbf{x} = \mathbf{o}, \boldsymbol{\theta}) \tag{12}$$

where $\alpha : \mathcal{P} \to \mathbb{Z}^+$ is the arity function for each predicate symbol, and ϕ and ψ are subformulas of φ. $\hat{p}$ computes the fuzzy degree of truth of some formula φ using the product norm and the Reichenbach implication [1], which makes our approach a special case of Real Logic [26] that we call *Product Real Logic*. The $\forall$ quantifier is interpreted in Equation 7 by going through all instantiations, which in this case is all n-tuples in the domain D_i, and also looping over all domains D_i in the set of domains (i.e., pictures) $\mathcal{O}_i$.

Example 1. The loss term associated with the formula $\varphi = \forall x, y \; \mathsf{chair}(x) \wedge \mathsf{partOf}(y, x) \to \mathsf{cushion}(y) \vee \mathsf{armRest}(y)$ is computed as follows:

$$\mathcal{L}(\boldsymbol{\theta}; \varphi, \mathcal{O}_u) = - \sum_{D \in \mathcal{O}, o_1, o_2 \in D} 1 - f_{\mathsf{chair}}^{\boldsymbol{\theta}}(o_1) \cdot f_{\mathsf{partOf}}^{\boldsymbol{\theta}}(o_2, o_1) \cdot$$

$$(1 - f_{\mathsf{cushion}}^{\boldsymbol{\theta}}(o_2)) \cdot (1 - f_{\mathsf{armRest}}^{\boldsymbol{\theta}}(o_2))$$

Say $\mathcal{O}_u$ contains the picture in Figure 2 whose domain is $\{a, b\}$ and the model predicts the following distribution over worlds:

$$f_{\mathsf{chair}}^{\boldsymbol{\theta}}(a) = 0.9 \qquad\qquad f_{\mathsf{chair}}^{\boldsymbol{\theta}}(b) = 0.4$$
$$f_{\mathsf{cushion}}^{\boldsymbol{\theta}}(a) = 0.05 \qquad\qquad f_{\mathsf{cushion}}^{\boldsymbol{\theta}}(b) = 0.5$$
$$f_{\mathsf{armRest}}^{\boldsymbol{\theta}}(a) = 0.05 \qquad\qquad f_{\mathsf{armRest}}^{\boldsymbol{\theta}}(b) = 0.1$$
$$f_{\mathsf{partOf}}^{\boldsymbol{\theta}}(a, a) = 0.001 \qquad\qquad f_{\mathsf{partOf}}^{\boldsymbol{\theta}}(b, b) = 0.001$$
$$f_{\mathsf{partOf}}^{\boldsymbol{\theta}}(a, b) = 0.01 \qquad\qquad f_{\mathsf{partOf}}^{\boldsymbol{\theta}}(b, a) = 0.95$$

The model returns high values for $f_{\mathsf{chair}}^{\boldsymbol{\theta}}(a)$ and $f_{\mathsf{partOf}}^{\boldsymbol{\theta}}(b, a)$ but it is not confident of $f_{\mathsf{cushion}}^{\boldsymbol{\theta}}(b)$, even though it is clearly higher than $f_{\mathsf{armRest}}^{\boldsymbol{\theta}}(b)$. We can decrease

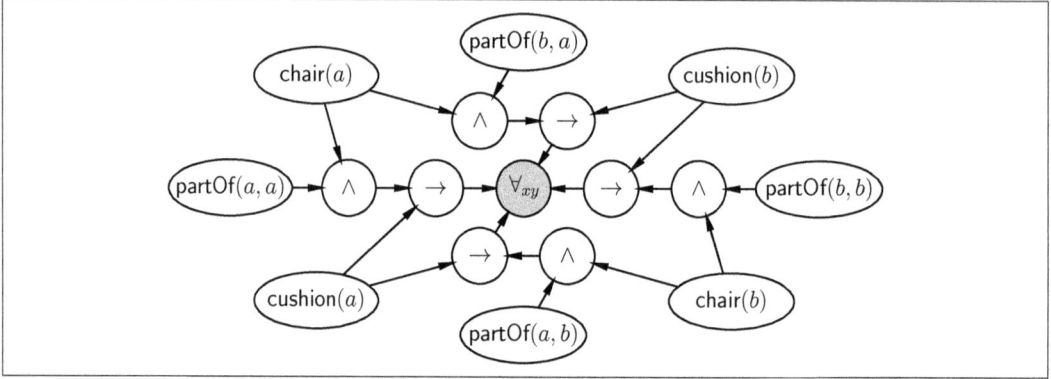

Figure 3: The Bayesian network associated with grounding of the formula $\forall x, y\ \mathsf{chair}(x) \wedge \mathsf{partOf}(y, x) \rightarrow \mathsf{cushion}(y)$ on the domain from Figure 2. We treat connectives and quantifiers as binary random variables (which correspond to subformulas through their parents) of which the conditional probabilities are computed using truth tables.

$\mathcal{L}(\boldsymbol{\theta}; \varphi, \mathcal{O}_u) = 0.612$ simply by increasing $f^{\boldsymbol{\theta}}_{\mathsf{cushion}}(b)$, since $f^{\boldsymbol{\theta}}_{\mathsf{cushion}}$ is a differentiable function with respect to $\boldsymbol{\theta}$.

This example shows that we can find a new instance of the cushion predicate using reasoning on an unlabeled dataset. This process uses both statistical reasoning and symbolic rules. As more data improves generalization, those additional examples could help reducing the sparsity of the SII problem. Furthermore, [7] showed that it is also possible to correct wrong labels due to noisy data when these do not satisfy the formulas.

Figure 3 shows the Bayesian Network for this formula on the picture from Figure 2, illustrating the computation path. We treat each subformula as a binary random variable of which the conditional probabilities are given by truth tables. Because the graph is not acyclic, we can use loopy belief propagation which is empirically shown to often be a good approximation of the correct probability [18]. In fact, Product Real Logic can be seen as performing a single iteration of belief propagation. However, this can be problematic. For example, the degree of truth of the ground formula $\mathsf{chair}(o) \wedge \mathsf{chair}(o)$ would be computed using $f^{\boldsymbol{\theta}}_{\mathsf{chair}}(o)^2$ instead of the probability of this statement, $f^{\boldsymbol{\theta}}_{\mathsf{chair}}(o)$ [22]. We show in Appendix A that Product Real Logic computes the correct probability $p(\mathcal{K}|\mathcal{O}_u, \boldsymbol{\theta})$ for a corpus $\mathcal{K}$ under the strong assumption that, after grounding, each ground atom is used at most once.

An interesting and useful property of our approach is that it can perform multi-hop reasoning in an iterative, yet extremely noisy, manner. In one iteration it might, for instance, increase $f^{\theta}_{\text{cushion}}(o)$. And since $f^{\theta}_{\text{cushion}}(o)$ will return higher values in future iterations, it can be used to prove that the probability of other ground atoms that occur in formulas with cushion(o) should also be increased or decreased.

A convenient property of the SII task is that we consider just binary relations between objects appearing on the same pictures. The Herbrand base then contains $O(|\mathcal{P}| \cdot |D_i|^2)$ ground atoms, which is feasible as there are often not more than a few dozen objects on an image. This property also holds in natural language to some degree in the following way: only the words appearing in the same paragraph can be related. This is in contrast to the knowledge base completion task where we have a single graph with many objects and predicates [27].

Figure 2: We can deduce that b is a cushion if we are confident about the truth value of chair(a) and partOf(b, a) using the formula $\forall x, y$ chair$(x) \wedge$ partOf$(y, x) \rightarrow$ cushion(y).

2.3 Implementation

We optimize the negative logarithm of the likelihood function given in Equation 4. In particular, we use minibatch gradient descent to decrease the computation time both for the supervised part of the loss and the unsupervised part. We turn the unsupervised loss into minibatch gradient descent by approximating the computation of the $\forall$ quantifier: instead of summing over all n-tuples and all domains, we randomly sample from these n-tuples independently from the domain it belongs to.

2.4 The Material Implication

To provide a better understanding of the inner machinery of our approach, we will elaborate on some interesting partial derivatives. Say, we have a formula φ of the form $\forall \mathbf{x} \phi(\mathbf{x}) \rightarrow \psi(\mathbf{x})$, where $\phi(\mathbf{x})$ is the antecedent and $\psi(\mathbf{x})$ the consequent of φ. First, we write out the partial derivative of $\mathcal{L}(\boldsymbol{\theta}; \varphi, \mathcal{O}_u)$ with respect to the consequent, where we make use of the chain rule:

$$d_\varphi^{\text{MP}}(\mathbf{o}) := \frac{\partial \log \hat{p}(\varphi | \mathcal{O}_u, \boldsymbol{\theta})}{\partial \hat{p}(\psi | \mathbf{o}, \boldsymbol{\theta})} = \frac{\partial \sum_{\mathbf{o} \in D, D \in \mathcal{O}_u} \log \hat{p}(\phi \to \psi | \mathbf{o}, \boldsymbol{\theta})}{\partial \hat{p}(\psi | \mathbf{o}, \boldsymbol{\theta})} \tag{13}$$

$$= \frac{\partial \sum_{\mathbf{o} \in D, D \in \mathcal{O}_u} \log(1 - \hat{p}(\phi | \mathbf{o}, \boldsymbol{\theta}) \cdot (1 - \hat{p}(\psi | \mathbf{o}, \boldsymbol{\theta})))}{\partial \hat{p}(\psi | \mathbf{o}, \boldsymbol{\theta})} \tag{14}$$

$$= \frac{\hat{p}(\phi | \mathbf{o}, \boldsymbol{\theta})}{1 - \hat{p}(\phi | \mathbf{o}, \boldsymbol{\theta}) \cdot (1 - \hat{p}(\psi | \mathbf{o}, \boldsymbol{\theta})))} = \frac{\hat{p}(\phi | \mathbf{o}, \boldsymbol{\theta})}{\hat{p}(\phi \to \psi | \mathbf{o}, \boldsymbol{\theta})} \tag{15}$$

$d_\varphi^{\text{MP}}(\mathbf{o})$ mirrors the application of the Modus Ponens (MP) rule using the implication $\phi \to \psi$ for the assignment of $\mathbf{o}$ to $\mathbf{x}$. The MP rule says that if ϕ is true and $\phi \to \psi$, then ψ should also be true. Similarly, if $\phi(\mathbf{o})$ is likely and $\phi \to \psi$, then $\psi(\mathbf{o})$ should also be likely. Indeed, notice that $d_\varphi^{\text{MP}}(\mathbf{o})$ grows with $\hat{p}(\phi | \mathbf{o}, \boldsymbol{\theta})$. Also, $d_\varphi^{\text{MP}}(\mathbf{o})$ is largest when $\hat{p}(\phi | \mathbf{o}, \boldsymbol{\theta})$ is high and $\hat{p}(\psi | \mathbf{o}, \boldsymbol{\theta})$ is low as it then approaches a singularity in the divisor. We next show the derivation with respect to the negated antecedent:

$$d_\varphi^{\text{MT}}(\mathbf{o}) := \frac{\partial \log \hat{p}(\varphi | \mathcal{O}_u, \boldsymbol{\theta})}{\partial \hat{p}(\neg \phi | \mathbf{o}, \boldsymbol{\theta})} = \frac{\hat{p}(\neg \psi | \mathbf{o}, \boldsymbol{\theta})}{\hat{p}(\phi \to \psi | \mathbf{o}, \boldsymbol{\theta})} \tag{16}$$

Similarly, it mirrors the application of the Modus Tollens (MT) rule which says that if ψ is false and $\phi \to \psi$, then ϕ should also be false. Again, realize that $d_\varphi^{\text{MT}}(\mathbf{o})$ grows with $\hat{p}(\psi | \mathbf{o}, \boldsymbol{\theta})$.

It is easy to see that $d_\varphi^{\text{MP}}(\mathbf{o}) > d_\varphi^{\text{MT}}(\mathbf{o})$ whenever $\hat{p}(\phi | \mathbf{o}, \boldsymbol{\theta}) > \hat{p}(\neg \psi | \mathbf{o}, \boldsymbol{\theta})$. Furthermore, the global minimum of $\mathcal{L}(\boldsymbol{\theta}; \varphi, \mathcal{O}_u)$ is some parameter value $\boldsymbol{\theta}^*$ so that $\hat{p}(\phi | \mathcal{O}_u, \boldsymbol{\theta}^*) = 0$ and $\hat{p}(\psi | \mathcal{O}_u, \boldsymbol{\theta}^*) = 1$ for all $\mathbf{o}$, which corresponds to the material implication.

Next, we show how these quantities are used in the updating of the parameters $\boldsymbol{\theta}$ using backpropagation and act as mixing components on the gradient updates:

$$\frac{\log \hat{p}(\varphi | \mathcal{O}_u, \boldsymbol{\theta})}{\partial \boldsymbol{\theta}} = \sum_{\mathbf{o} \in D, D \in \mathcal{O}_u} d_\varphi^{\text{MP}}(\mathbf{o}) \cdot \frac{\partial \hat{p}(\psi | \mathbf{o}, \boldsymbol{\theta})}{\partial \boldsymbol{\theta}} + d_\varphi^{\text{MT}}(\mathbf{o}) \cdot \frac{\partial \hat{p}(\neg \phi | \mathbf{o}, \boldsymbol{\theta})}{\partial \boldsymbol{\theta}} \tag{17}$$

2.5 The Raven Paradox

In our experiments, we have found that this approach is very sensitive to the *raven paradox* [10]. It is stated as follows: Assuming that observing an example of a statement is evidence for that statement (i.e., the degree of belief in that statement increases), and that evidence for a sentence also is evidence for all the other logically

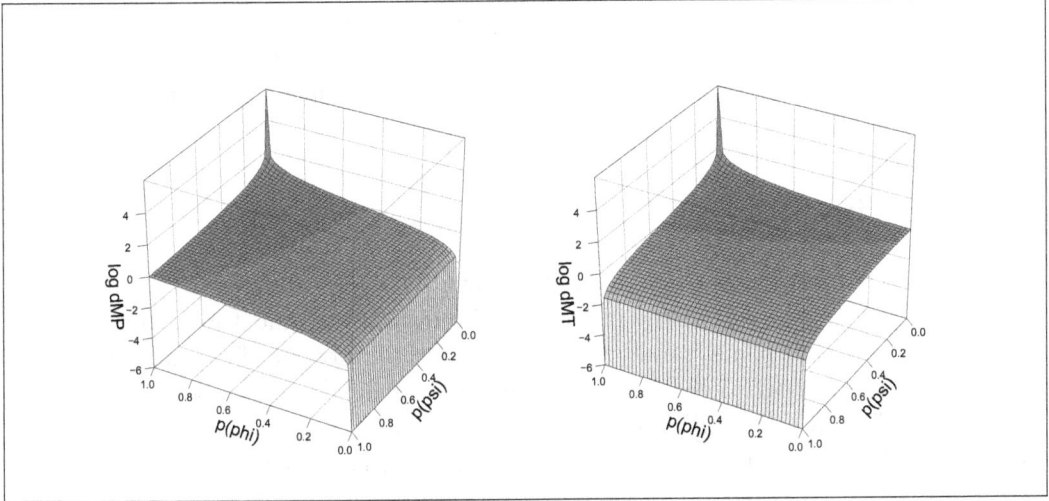

Figure 4: Plots of $d_\varphi^{\mathrm{MP}}(\mathbf{o})$ (Equation 15) and $d_\varphi^{\mathrm{MT}}(\mathbf{o})$ (Equation 16). Note that the y axis is using a log scale.

equivalent sentences, then our belief in "ravens are black" increases when we observe non-black non-raven, by the contrapositive "non-ravens are non-black". Equation 17 shows however that the gradient is equally determined by positive evidence (observing black ravens) as by contrapositive evidence (observing non-black non-ravens). Because in the real world there are far more ravens than non-black objects, optimizing $\hat{p}(\forall o\ \mathsf{raven}(o) \to \mathsf{black}(o)|\mathcal{O}_u, \boldsymbol{\theta})$ amounts to recognizing that something is not a raven when it is not black. However, Machine Learning models tend to be biased when the class distribution is unbalanced during training [30].

Figure 4 shows plots of $d_\varphi^{\mathrm{MP}}(\mathbf{o})$ and $d_\varphi^{\mathrm{MT}}(\mathbf{o})$ for different values of $\hat{p}(\phi|\mathbf{o}, \boldsymbol{\theta})$ and $\hat{p}(\psi|\mathbf{o}, \boldsymbol{\theta})$. In practice, for many formulas of this form, the most common case will be that the model predicts $\neg\phi(\mathbf{o}) \wedge \neg\psi(\mathbf{o})$. Then, $d_\varphi^{\mathrm{MP}}(\mathbf{o})$ approaches 0 and $d_\varphi^{\mathrm{MT}}(\mathbf{o})$ will be around 1. For instance, the average value of $d_\varphi^{\mathrm{MP}}(\mathbf{o})$ for the problem in Example 1 is 0.214, while the average value of $d_\varphi^{\mathrm{MT}}(\mathbf{o})$ is 0.458.

We analyze a naive way of dealing with this phenomenon. We normalize the contribution to the total gradient of MP and MT reasoning by replacing the loss function $\mathcal{L}$ of rules of the form $\forall\mathbf{x}\phi(\mathbf{x}) \to \psi(\mathbf{x})$ as follows:

$$\mathcal{L}(\boldsymbol{\theta}; \mathcal{K}, \mathcal{O}_u) = -\sum_{\varphi \in \mathcal{K}} \sum_{\mathbf{o} \in D, D \in \mathcal{O}_u} \frac{\mu \cdot d_\varphi^{\mathrm{MP}}(\mathbf{o})}{\sum_{\mathbf{o}' \in D, D \in \mathcal{O}_u} d_\varphi^{\mathrm{MP}}(\mathbf{o}')} \cdot \hat{p}(\psi|\mathbf{o}, \boldsymbol{\theta})$$
$$+ \frac{(1-\mu) \cdot d_\varphi^{\mathrm{MT}}(\mathbf{o})}{\sum_{\mathbf{o}' \in D, D \in \mathcal{O}_u} d_\varphi^{\mathrm{MT}}(\mathbf{o}')} \cdot \hat{p}(\neg\phi|\mathbf{o}, \boldsymbol{\theta})$$
(18)

where μ is a hyperparameter that assigns the relative importance of Modus Ponens with respect to Modus Tollens updates. We are then able to control how much either contributes to the training process. We experiment with different values of μ and report our findings in the next section.

3   Experiments

We carried out simple experiments on the PASCAL-PART dataset [3] in which the task is to predict the type of the object in a bounding box and the partOf relation which expresses that some bounding box is a part of another. For example, a tail can be a part of a cat. Like in [7], the output softmax layer over the 64 object classes of a Fast R-CNN [9] detector is used for the bounding box features. Note that this makes the problem of recognizing types very easy as the features correlate strongly with the true output types. Therefore, to get a more realistic estimate, we randomly split the dataset into only 7 labeled pictures for $\mathcal{D}_l$ and 2128 unlabeled pictures for $\mathcal{D}_u$. Additionally, we only consider 11 (related) types out of 64 due to computational constraints. As there is a large amount of variance associated with randomly splitting in this way, we do all our experiments on 20 random splits of the dataset. The results are evaluated on a held-out validation set of 200 images. We compare the accuracy of prediction of the type of the bounding box and the AUC (area under curve) for the partOf relationship.

We model $f^{\theta}_{\mathsf{type}_i}(o)$ using a single Logic Tensor Network (LTN) layer [7] of width 10 followed by a softmax output layer to ensure mutual exclusivity of types. The term $f^{\theta}_{\mathsf{partOf}}(o_1, o_2)$ is modeled using an LTN layer of width 2 and a sigmoid output layer. The loss function is then optimized using RMSProp over 6000 iterations. We use the same relational background knowledge as [7] which are rules like the following:

$$\forall x, y \; \mathsf{chair}(x) \wedge \mathsf{partOf}(y, x) \to \mathsf{cushion}(y) \vee \mathsf{armRest}(y)$$
$$\forall x, y \; \mathsf{cushion}(x) \wedge \mathsf{partOf}(x, y) \to \mathsf{chair}(y) \vee \mathsf{bench}(y)$$
$$\forall x \; \neg\mathsf{partOf}(x, x)$$
$$\forall x, y \; \mathsf{partOf}(x, y) \to \neg\mathsf{partOf}(y, x)$$

We compare three methods. In the first one we train without any rules, which forms the *supervised* baseline. In the second, *unnormalized*, we add the rules to the unlabeled data. This does not use any technique for dealing with the raven paradox. In the last one called *normalized*, we normalize MP and MT reasoning using Equation 18 for several different values of μ. The results in Table 1 are statistically significant when using a paired t-test.

	Precision types
Supervised	0.440 ± 0.0013
Unnormalized	0.455 ± 0.0014
Normalized $\mu = 0$	0.454 ± 0.0015
Normalized $\mu = 0.1$	0.505 ± 0.0014
Normalized $\mu = 0.25$	$\mathbf{0.517 \pm 0.0013}$
Normalized $\mu = 0.5$	0.510 ± 0.0013
Normalized $\mu = 0.75$	0.496 ± 0.0012
Normalized $\mu = 1$	0.435 ± 0.0015

Table 1: Results of the experiments. 20 runs using random splits of the data are averaged alongside 95% confidence intervals. All results are significant.

4 Analysis

Our experiments show that we can significantly improve on the classification of the types of objects for this problem. The normalized method in particular outperforms the unnormalized method, suggesting that explicitly dealing with the raven paradox is essential in this problem.

4.1 Gradient Updates

We analyze how the different methods handle the implication using the quantities d_φ^{MP} and d_φ^{MT} defined in Section 2.4. Figure 5 shows the average magnitude of d_φ^{MP} and d_φ^{MT} in the unnormalized model, which is computed by averaging over all training examples and formulas. This shows that the average MT gradient update is, in this problem, around 100 times larger than the average MP gradient update, i.e., it uses far more contrapositive reasoning. The unnormalized method acts very similar to the normalized one with $\mu \approx 0.01$.

Next, we will analyze how accurate our approach is at reasoning by comparing its 'decisions' to what should have been the correct 'decision'. We sample 2000 pairs of bounding boxes from the PASCAL-PART test set $\langle \mathcal{O}_t, \mathcal{W}_t \rangle$. We consider a pair of bounding boxes $\mathbf{o}$ from an image i in the test set $\mathcal{O}_t$. $d_{\varphi}^{\mathrm{MP}}(\mathbf{o})$ is a *correctly reasoned gradient* if both $\phi(\mathbf{o})$ and $\psi(\mathbf{o})$ are true in $\mathbf{w}_i^t$. Likewise, $d_{\varphi}^{\mathrm{MT}}(\mathbf{o})$ is a correctly reasoned gradient if $\neg\psi(\mathbf{o})$ and $\neg\phi(\mathbf{o})$ are true in $\mathbf{w}_i^t$. Furthermore, we say that $d_{\varphi}^{\mathrm{MP}}(\mathbf{o})$ is a *correctly updated gradient* if at least $\psi(\mathbf{o})$ is true in $\mathbf{w}_i^t$, and that $d_{\varphi}^{\mathrm{MT}}(\mathbf{o})$ is

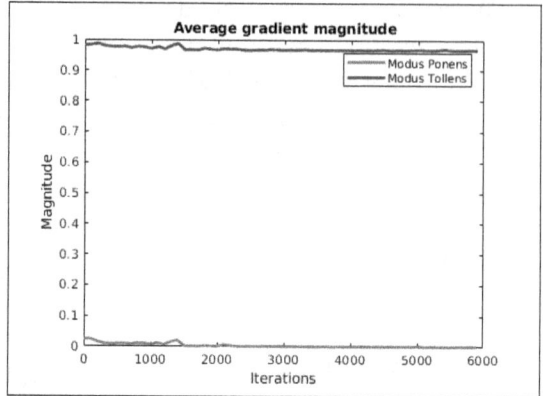

Figure 5: The average magnitude of Modus Ponens and Modus Tollens gradients.

correctly updated when $\neg\phi(\mathbf{o})$ is true in $\mathbf{w}_i^t$. Then the *correctly reasoned ratios* are computed using

$$\mathrm{cr}^{\mathrm{MP}} = \frac{\sum_{\varphi \in \mathcal{K}} \sum_{\mathbf{o} \in D_i, D_i \in \mathcal{O}_t} v(\phi, \mathbf{w}_i^t) \cdot v(\psi, \mathbf{w}_i^t) \cdot d_{\varphi}^{\mathrm{MP}}(\mathbf{o})}{\sum_{\varphi \in \mathcal{K}} \sum_{\mathbf{o}' \in D_i, D_i \in \mathcal{O}_t} d_{\varphi}^{\mathrm{MP}}(\mathbf{o}')} \tag{19}$$

$$\mathrm{cr}^{\mathrm{MT}} = \frac{\sum_{\varphi \in \mathcal{K}} \sum_{\mathbf{o} \in D_i, D_i \in \mathcal{O}_t} v(\neg\phi, \mathbf{w}_i^t) \cdot v(\neg\psi, \mathbf{w}_i^t) \cdot d_{\varphi}^{\mathrm{MT}}(\mathbf{o})}{\sum_{\varphi \in \mathcal{K}} \sum_{\mathbf{o}' \in D_i, D_i \in \mathcal{O}_t} d_{\varphi}^{\mathrm{MT}}(\mathbf{o}')}. \tag{20}$$

The definition of the *correctly updated ratios* ($\mathrm{cu}^{\mathrm{MP}}$ and $\mathrm{cu}^{\mathrm{MT}}$) are nearly the same. $\mathrm{cu}^{\mathrm{MP}}$ is found by removing the $v(\phi, \mathbf{w}_i^t)$ term from Equation 19, and $\mathrm{cu}^{\mathrm{MT}}$ by removing the $v(\neg\psi, \mathbf{w}_i^t)$ term from Equation 20.

Figure 6 shows the value of these ratios during training. The dotted lines that represent MT reasoning shows a convenient property, namely that is nearly always correct because of the large class imbalance. This could be the reason there is a significant benefit to adding contrapositive reasoning. Both normalized and unnormalized at $\mu = 1$ seems to get 'better' at reasoning during training, as the correctly updated ratios go up. After training for some time, the unnormalized method seems to be best at reasoning correctly for both MP and MT. Another interesting observation is the difference between $\mathrm{cr}^{\mathrm{MP}}$ and $\mathrm{cu}^{\mathrm{MP}}$. At many points, about half of the gradient magnitude correctly increases $\hat{p}(\psi|\mathbf{o}, \boldsymbol{\theta})$ because the model predicts a high value for $\hat{p}(\phi|\mathbf{o}, \boldsymbol{\theta})$, even though $\phi(\mathbf{o})$ is not actually in the test labels. It is interesting to see that, this kind of faulty reasoning which does lead to the right conclusion is actually beneficial for training.

Furthermore, disabling MT completely by setting μ to 1 seems to destabilize the

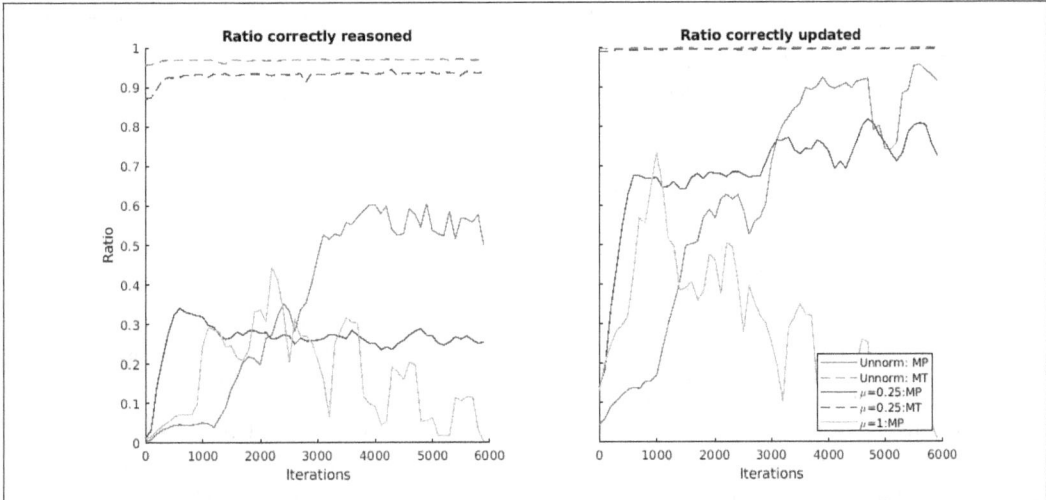

Figure 6: The left plot shows cr^{MP} and cr^{MT} and the right plot cu^{MP} and cu^{MT} for the Unnormalized method (denoted as Unnorm) and the Normalized methods with $\mu = 0.25$ and $\mu = 1$.

reasoning. This is also reflected in the validation accuracy that seems to decline when cu^{MP} declines. This suggests that contrapositive reasoning is required to increase the amount of correct gradient updates.

5 Related work

5.1 Injecting Logic into Parameterized Models

Our work follows the recent works on Real Logic [26, 7], and the method we use is a special case of Real Logic with some additional changes. A particular difference is that the logic we employ has no function symbols, which was due to simplicity purposes. Injecting background knowledge into vector embeddings of entities and relations has been studied in [5, 6, 20, 21]. In particular, [22] has some similarities with Real Logic and our method. However, this method is developed for regularizing vector embeddings instead of any parameterized model. In this sense, it can also be seen as a special case of Real Logic. Semantic Loss [31] is a very similar semi-supervised learning method. This loss is essentially Equation 4, which makes it more accurate than Product Real Logic, but also exponential in runtime. To deal with this, they compile SDD's [4] to make the computation tractable. A recent direction is DeepProbLog [17], a probabilistic version of Prolog with neural predicates that also uses SDD's. [11] also injects rules into a general model with a framework that transfers the logic rules using a so-called teacher network. This model is significantly different from the aforementioned ones, as it does not add a loss for each rule.

5.2 Semi-Supervised Learning

There is a large body of literature on semi-supervised methods [19, 2]. In particular, recent research on graph-based semi-supervised learning [14, 32, 33] relates unlabeled and labeled data through a graph structure. However, they do not use logically structured background knowledge. It is generally used for entity classification, although in [25] it is also used on link prediction. [16] introduced the surprisingly effective method Pseudo-Label that first trains a model using the labeled dataset, then labels the unlabeled dataset using this model and continues training on this newly labeled dataset. Our approach has a similar intuition in that we use the current model to get an estimation about the correct labels of the labeled dataset, and then use those labels to predict remaining labels, but the difference is that we use background knowledge to choose these labels.

6 Conclusion and Future Work

We proposed a novel semi-supervised learning technique and showed that it is possible to find labels for samples in an unlabeled dataset by evaluating them on relational background knowledge. Since implication is at the core of logical reasoning, we analyzed this by inspecting the gradients with respect to the antecedent and the consequent. Surprisingly, we discovered a strong imbalance between the contributions of updates from MP and MT in the induction process. It turned out that our approach is highly sensitive to the Raven paradox [10] requiring us to handle positive and contrapositive reasoning separately. Normalizing these different types of reasoning yields the largest improvements to the supervised baseline. Since it is quite general, we suspect that issues with this imbalance could occur in many systems that perform inductive reasoning.

We would like to investigate this phenomenon with different background knowledge and different datasets such as VisualGenome and ImageNet. In particular, we are interested in other approaches for modelling the implication like different Fuzzy Implications [12] or by taking inspiration from Bayesian treatments of the Raven paradox [29]. Furthermore, it could be applied to natural language understanding tasks like semantic parsing.

References

[1] Merrie Bergmann. *An introduction to many-valued and fuzzy logic: semantics, algebras, and derivation systems.* Cambridge University Press, 2008.

[2] Olivier Chapelle, Bernhard Scholkopf, and Alexander Zien. *Semi-supervised learning.* 2006.

[3] Xianjie Chen, Roozbeh Mottaghi, Xiaobai Liu, Sanja Fidler, Raquel Urtasun, and Alan Yuille. Detect what you can: Detecting and representing objects using holistic models and body parts. In *Proceedings of the IEEE Conference on Computer Vision and Pattern Recognition*, pages 1971–1978, 2014.

[4] Adnan Darwiche. SDD: A new canonical representation of propositional knowledge bases. *IJCAI International Joint Conference on Artificial Intelligence*, pages 819–826, 2011.

[5] Thomas Demeester, Tim Rocktäschel, and Sebastian Riedel. Lifted rule injection for relation embeddings. In *Proceedings of the 2016 Conference on Empirical Methods in Natural Language Processing*, pages 1389–1399. Association for Computational Linguistics, 2016.

[6] Thomas Demeester, Tim Rocktäschel, and Sebastian Riedel. Regularizing relation representations by first-order implications. In *AKBC2016, the Workshop on Automated Base Construction*, pages 1–6, 2016.

[7] Ivan Donadello, Luciano Serafini, and Artur S. d'Avila Garcez. Logic tensor networks for semantic image interpretation. In *IJCAI*, pages 1596–1602. ijcai.org, 2017.

[8] Dorothy Edgington. Indicative conditionals. In Edward N. Zalta, editor, *The Stanford Encyclopedia of Philosophy*. Metaphysics Research Lab, Stanford University, winter 2014 edition, 2014.

[9] Ross Girshick. Fast r-cnn. *International Conference on Computer Vision*, pages 1440–1448, 2015.

[10] Carl G Hempel. Studies in the logic of confirmation (i.). *Mind*, 54(213):1–26, 1945.

[11] Zhiting Hu, Xuezhe Ma, Zhengzhong Liu, Eduard Hovy, and Eric Xing. Harnessing deep neural networks with logic rules. In *Proceedings of the 54th Annual Meeting of the Association for Computational Linguistics (Volume 1: Long Papers)*, pages 2410–2420. Association for Computational Linguistics, 2016.

[12] Balasubramaniam Jayaram and Michal Baczynski. *Fuzzy Implications*, volume 231. 2008.

[13] Justin Johnson, Ranjay Krishna, Michael Stark, Li-Jia Li, David Shamma, Michael Bernstein, and Li Fei-Fei. Image retrieval using scene graphs. In *Proceedings of the IEEE conference on computer vision and pattern recognition*, pages 3668–3678, 2015.

[14] Thomas N Kipf and Max Welling. Semi-supervised classification with graph convolutional networks. 2016.

[15] Ranjay Krishna, Yuke Zhu, Oliver Groth, Justin Johnson, Kenji Hata, Joshua Kravitz, Stephanie Chen, Yannis Kalantidis, Li-Jia Li, David A Shamma, et al. Visual genome: Connecting language and vision using crowdsourced dense image annotations. *International Journal of Computer Vision*, 123(1):32–73, 2017.

[16] Dong-Hyun Lee. Pseudo-label: The simple and efficient semi-supervised learning method for deep neural networks. 2013.

[17] Robin Manhaeve, Sebastijan Dumančić, Angelika Kimmig, Thomas Demeester, and Luc De Raedt. Deepproblog: Neural probabilistic logic programming. *arXiv preprint arXiv:1805.10872*, 2018.

[18] Kevin Murphy, Yair Weiss, and Michael I. Jordan. Loopy Belief Propagation for Approximate Inference: An Empirical Study. pages 467–476, 2013.

[19] Avital Oliver, Augustus Odena, Colin Raffel, Ekin D. Cubuk, and Ian J. Goodfellow. Realistic evaluation of semi-supervised learning algorithms. 2018.

[20] Tim Rocktäschel. Combining representation learning with logic for language processing. *CoRR*, abs/1712.09687, 2017.

[21] Tim Rocktäschel and Sebastian Riedel. End-to-end differentiable proving. In *Advances in Neural Information Processing Systems*, pages 3791–3803, 2017.

[22] Tim Rocktäschel, Sameer Singh, and Sebastian Riedel. Injecting logical background knowledge into embeddings for relation extraction. In *Proceedings of the 2015 Conference of the North American Chapter of the Association for Computational Linguistics: Human Language Technologies*, pages 1119–1129, 2015.

[23] Dan Roth. On the hardness of approximate reasoning. *Artificial Intelligence*, 82(1-2):273–302, 1996.

[24] Olga Russakovsky, Jia Deng, Hao Su, Jonathan Krause, Sanjeev Satheesh, Sean Ma, Zhiheng Huang, Andrej Karpathy, Aditya Khosla, Michael Bernstein, Alexander C. Berg, and Li Fei-Fei. ImageNet Large Scale Visual Recognition Challenge. *International Journal of Computer Vision (IJCV)*, 115(3):211–252, 2015.

[25] Michael Schlichtkrull, Thomas N Kipf, Peter Bloem, Rianne van den Berg, Ivan Titov, and Max Welling. Modeling Relational Data with Graph Convolutional Networks. In Aldo Gangemi, Roberto Navigli, Maria-Esther Vidal, Pascal Hitzler, Raphaël Troncy, Laura Hollink, Anna Tordai, and Mehwish Alam, editors, *The Semantic Web*, pages 593–607, Cham, 2018. Springer International Publishing.

[26] Luciano Serafini and Artur d'Avila Garcez. Logic tensor networks: Deep learning and logical reasoning from data and knowledge. 2016.

[27] Richard Socher, Danqi Chen, Christopher D Manning, and Andrew Ng. Reasoning with neural tensor networks for knowledge base completion. In *Advances in neural information processing systems*, pages 926–934, 2013.

[28] Dirk Van Dalen. *Logic and structure*. Springer, 2004.

[29] Peter B.M. Vranas. Hempel's raven paradox: A lacuna in the standard Bayesian solution. *British Journal for the Philosophy of Science*, 55(3):545–560, 2004.

[30] Gm Weiss and Foster Provost. The effect of class distribution on classifier learning: an empirical study. *Rutgers University*, (September 2001), 2001.

[31] Jingyi Xu, Zilu Zhang, Tal Friedman, Yitao Liang, and Guy Van den Broeck. A semantic loss function for deep learning with symbolic knowledge. In Jennifer Dy and Andreas Krause, editors, *Proceedings of the 35th International Conference on Machine*

Learning, volume 80 of *Proceedings of Machine Learning Research*, pages 5502–5511, Stockholmsmässan, Stockholm Sweden, 10–15 Jul 2018. PMLR.

[32] Zhilin Yang, William W. Cohen, and Ruslan Salakhutdinov. Revisiting semi-supervised learning with graph embeddings. In *Proceedings of the 33rd International Conference on International Conference on Machine Learning - Volume 48*, ICML'16, pages 40–48. JMLR.org, 2016.

[33] Xiaojin Zhu, Zoubin Ghahramani, and John D Lafferty. Semi-supervised learning using gaussian fields and harmonic functions. In *Proceedings of the 20th International conference on Machine learning (ICML-03)*, pages 912–919, 2003.

Appendices

A   Conditional Optimality of Product Real Logic

Considering only a single domain D of objects $x \in \mathbb{R}^D$, we have the Herbrand base $\mathbf{A}$. Let $\varphi \in \mathcal{K}$ be a set of function-free FOL formulas in Skolem-normal form. Furthermore, let $\mathcal{P} = \{\mathsf{P}_1, ..., \mathsf{P}_K\}$ be a set of predicates which for ease of notation and without loss of generality we assume to all have the arity α.

Each ground atom $\mathsf{P}(\mathbf{o}) \sim \mathrm{Bern}(f_\mathsf{P}^{\boldsymbol{\theta}}(\mathbf{o}))$ is a binary random variable that denotes the binary truth value. It is distributed by a Bernoulli distribution with mean $f_\mathsf{P}^{\boldsymbol{\theta}} \in \mathbb{R}^{\alpha \times D} \to [0,1]$.

For each formula φ, we have the set of ground atoms $\mathbf{A}_\varphi \subseteq \mathbf{A}$ appearing in the instantiations of φ. Likewise, the assignment of truth values of $\mathbf{A}_\varphi$ is $\mathbf{w}_\varphi$, which is a subset of the world $\mathbf{w}$. We can now express the joint probability, using Equation 1 and the valuation function defined in Section 2.1:

$$p(\mathcal{K}, \mathbf{w}|D, \boldsymbol{\theta}) = p(\mathbf{w}|\boldsymbol{\theta}) \cdot \prod_{\phi \in \mathcal{K}} v(\phi, \mathbf{w}_\varphi) \tag{21}$$

We will first show that Product Real Logic is equal to this probability with two strong assumptions. The first is that the sets of ground atoms $\mathbf{A}_\varphi$ are disjoint for all formulas in the corpus, i.e. if

$$\bigcup_{\varphi \in \mathcal{K}} \mathbf{A}_\varphi = \emptyset \tag{22}$$

The second is that the set of ground atoms used in two children (a direct subformula) of some subformula of a formula in $\mathcal{K}$ are disjoint. If $pa(\phi)$ returns the parent of ϕ and $r(\phi)$ returns the root of ϕ (the formula highest up the tree), then

$$\mathbf{A}_\phi \cup \mathbf{A}_\psi = \emptyset, \forall\{\phi, \psi|pa(\phi) = pa(\psi) \wedge r(\phi) \in \mathcal{K}\} \tag{23}$$

First, we marginalize over the different possible worlds:

$$p(\mathcal{K}|D,\boldsymbol{\theta}) = \sum_{\mathbf{w}} p(\mathbf{w}|\boldsymbol{\theta}) \cdot \prod_{\varphi \in \mathcal{K}} v(\phi, \mathbf{w}_\varphi) \tag{24}$$

$$= \sum_{\mathbf{w}_{\varphi_1}} p(\mathbf{w}_{\varphi_1}|\boldsymbol{\theta}) \cdot \left(\cdots \cdot \sum_{\mathbf{w}_{\varphi_{|\mathcal{K}|}}} p(\mathbf{w}_{\varphi_{|\mathcal{K}|}}|\boldsymbol{\theta}) \cdot \prod_{\varphi \in \mathcal{K}} v(\phi, \mathbf{w}_\varphi) \right) \tag{25}$$

$$= \sum_{\mathbf{w}_{\varphi_1}} p(\mathbf{w}_{\varphi_1}|\boldsymbol{\theta}) \cdot v(\varphi_1, \mathbf{w}_{\varphi_1}) \cdot \left(\cdots \cdot \sum_{\mathbf{w}_{\varphi_{|\mathcal{K}|}}} p(\mathbf{w}_{\varphi_{|\mathcal{K}|}}|\boldsymbol{\theta}) \cdot v(\varphi_{|\mathcal{K}|}, \mathbf{w}_{\varphi_{|\mathcal{K}|}}) \right) \tag{26}$$

$$= \prod_{\varphi \in \mathcal{K}} \sum_{\mathbf{w}_\varphi} p(\mathbf{w}_\varphi|\boldsymbol{\theta}) \cdot v(\phi, \mathbf{w}_\varphi) \tag{27}$$

where we make use of Equation 22 to join the summations, the independence of the probabilities of atoms from Equation 1 and marginalization of the atoms other than those in $\mathbf{A}_\varphi$.

We denote the set of instantiations of φ by S_φ, and a particular instance by s. $\mathbf{A}_s \subseteq \mathbf{A}_\varphi$ then is the set of ground atoms in s (and respectively for $\mathbf{w}_s$). Next we show that $\sum_{\mathbf{w}_\varphi} p(\mathbf{w}_\varphi|\boldsymbol{\theta}) \cdot v(\varphi, \mathbf{w}_\varphi) = \prod_{s \in S_\varphi} \hat{p}(\varphi|s, \boldsymbol{\theta})$. As the formulas are in prenex normal form, $\varphi = \forall x_1, ..., x_\alpha \phi$. We find that, using Equation 23 and the same procedure as in Equations 24-27

$$\sum_{\mathbf{w}_\varphi} p(\mathbf{w}_\varphi|\boldsymbol{\theta}) \cdot v(\varphi, \mathbf{w}_\varphi) = \sum_{\mathbf{w}_\varphi} p(\mathbf{w}_\varphi|\boldsymbol{\theta}) \cdot \prod_{s \in S_\varphi} v(\phi, \mathbf{w}_s) \tag{28}$$

$$= \prod_{s \in S_\varphi} \sum_{\mathbf{w}_s} p(\mathbf{w}_s|\boldsymbol{\theta}) \cdot v(\phi, \mathbf{w}_s). \tag{29}$$

Then, it suffices to show that $\sum_{\mathbf{w}_s} p(\mathbf{w}_s|\boldsymbol{\theta}) \cdot v(\phi, \mathbf{w}_s) = \hat{p}(\phi|s, \boldsymbol{\theta})$. This is done using recursion. For brevity, we will only proof it for the $\neg$ and $\wedge$ connectives, as we can proof the others using those.

Assume that $\phi = \mathsf{P}(x_1, ..., x_n)$. Then if $w_s(\mathsf{P}(x_1, ..., x_n))$ is the binary random variable of the ground atom $\mathsf{P}(x_1, ..., x_n)$ under the instantiation s,

$$\sum_{\mathbf{w}_s} p(\mathbf{w}_s|\boldsymbol{\theta}) \cdot v(\mathsf{P}(x_1, ..., x_n), \mathbf{w}_s) \tag{30}$$

$$= \sum_{\mathbf{w}_s \backslash \{w_s(\mathsf{P}(x_1,...,x_n))\}} p(\mathbf{w}_s \backslash \{w_s(\mathsf{P}(x_1, ..., x_n))\}|\boldsymbol{\theta}) \cdot \tag{31}$$

$$\sum_{w_s(\mathsf{P}(x_1,...,x_n))} p(w_s(\mathsf{P}(x_1, ..., x_n))|\boldsymbol{\theta}) \cdot w_s(\mathsf{P}(x_1, ..., x_n)) \tag{32}$$

$$= p(w_s(\mathsf{P}(x_1, ..., x_n))|\boldsymbol{\theta}) = \hat{p}(\mathsf{P}(x_1, ..., x_n)|s, \boldsymbol{\theta}). \tag{33}$$

Marginalize out all variables but $w_s(\mathsf{P}(x_1, ..., x_n))$. $v(\mathsf{P}(x_1, ..., x_n), \mathbf{w}_s)$ is 1 if $w_s(\mathsf{P}(x_1, ..., x_n))$ is, and 0 otherwise.

Next, assume $\phi = \neg\psi$. Then

$$\sum_{\mathbf{w}_s} p(\mathbf{w}_s|\boldsymbol{\theta}) \cdot v(\neg\psi, \mathbf{w}_s) \tag{34}$$

$$= \sum_{\mathbf{w}_s} p(\mathbf{w}_s|\boldsymbol{\theta}) \cdot (1 - v(\psi, \mathbf{w}_s)) \tag{35}$$

$$= \sum_{\mathbf{w}_s} p(\mathbf{w}_s|\boldsymbol{\theta}) - \sum_{\mathbf{w}_s} p(\mathbf{w}_s|\boldsymbol{\theta}) \cdot v(\psi, \mathbf{w}_s) \tag{36}$$

$$= 1 - \sum_{\mathbf{w}_s} p(\mathbf{w}_s|\boldsymbol{\theta}) \cdot v(\psi, \mathbf{w}_s) = \hat{p}(\neg\psi|s, \boldsymbol{\theta}) \tag{37}$$

Finally, assume $\varphi = \phi \wedge \psi$. Then

$$\sum_{\mathbf{w}_s} p(\mathbf{w}_s|\boldsymbol{\theta}) \cdot v(\phi \wedge \psi, \mathbf{w}_s) \tag{38}$$

$$= \sum_{\mathbf{w}_s} p(\mathbf{w}_s|\boldsymbol{\theta}) \cdot v(\phi, \mathbf{w}_s) \cdot v(\psi, \mathbf{w}_s) \tag{39}$$

$$= \sum_{\mathbf{w}_{\phi_s}} \sum_{\mathbf{w}_{\psi_s}} p(\mathbf{w}_{\phi_s}|\boldsymbol{\theta}) \cdot p(\mathbf{w}_{\psi_s}|\boldsymbol{\theta}) \cdot v(\phi, \mathbf{w}_{\phi_s}) \cdot v(\psi, \mathbf{w}_{\psi_s}) \cdot \tag{40}$$

$$\sum_{\mathbf{w}_s \setminus (\mathbf{w}_{\phi_s} \cup \mathbf{w}_{\psi_s})} p(\mathbf{w}_s \setminus (\mathbf{w}_{\phi_s} \cup \mathbf{w}_{\psi_s})|\boldsymbol{\theta}) \tag{41}$$

$$= \sum_{\mathbf{w}_{\phi_s}} p(\mathbf{w}_{\phi_s}|\boldsymbol{\theta}) \cdot v(\phi, \mathbf{w}_{\phi_s}) \cdot \sum_{\mathbf{w}_{\psi_s}} p(\mathbf{w}_{\psi_s}|\boldsymbol{\theta}) \cdot v(\psi, \mathbf{w}_{\psi_s}) \tag{42}$$

$$= \sum_{\mathbf{w}_s} p(\mathbf{w}_s|\boldsymbol{\theta}) \cdot v(\phi, \mathbf{w}_s) \cdot \sum_{\mathbf{w}_s} p(\mathbf{w}_s|\boldsymbol{\theta}) \cdot v(\psi, \mathbf{w}_s) \tag{43}$$

$$= \hat{p}(\phi|s, \boldsymbol{\theta}) \cdot \hat{p}(\psi|s, \boldsymbol{\theta}) = \hat{p}(\phi \wedge \psi|s, \boldsymbol{\theta}) \tag{44}$$

Using this result and equations 27 and 29, we find that

$$p(\mathcal{K}|D, \boldsymbol{\theta}) = \prod_{\varphi \in \mathcal{K}} \sum_{\mathbf{w}_\varphi} p(\mathbf{w}_\varphi|\boldsymbol{\theta}) \cdot v(\phi, \mathbf{w}_\varphi) \tag{45}$$

$$= \prod_{\varphi \in \mathcal{K}} \prod_{s \in S_\varphi} \sum_{\mathbf{w}_s} p(\mathbf{w}_s|\boldsymbol{\theta}) \cdot v(\phi, \mathbf{w}_s) \tag{46}$$

$$= \prod_{\varphi \in \mathcal{K}} \prod_{s \in S_\varphi} \hat{p}(\phi|s, \boldsymbol{\theta}) \tag{47}$$

 Received 18 June 2018

High-order Networks that Learn to Satisfy Logic Constraints

Gadi Pinkas and Shimon Cohen
Afeka Tel-Aviv Academic College of Engineering
Afeka Center for Language Processing

Abstract

Logic-based problems such as planning, formal verification and inference, typically involve combinatorial search and structured knowledge representation. Artificial neural networks (ANNs) are very successful statistical learners; however, they have been criticized for their weaknesses in representing and in processing complex structured knowledge which is crucial for combinatorial search and symbol manipulation. Two high-order neural architectures are presented (Symmetric and RNN), which can encode structured relational knowledge in neural activation, and store bounded First Order Logic (FOL) constraints in connection weights. Both architectures learn to search for a solution that satisfies the constraints. Learning is done by unsupervised "practicing" on problem instances from the same domain, in a way that improves the network-solving speed. No teacher exists to provide answers for the problem instances of the training and test sets. However, the domain constraints are provided as prior knowledge encoded in a loss function that measures the degree of constraint violations. Iterations of activation calculation and learning are executed until a solution that maximally satisfies the constraints emerges on the output units. As a test case, block-world planning problems are used to train flat networks with high-order connections that learn to plan in that domain, but the techniques proposed could be used more generally as in integrating prior symbolic knowledge with statistical learning.

Keywords: artificial neural networks, planning as SAT, constraint satisfaction, unsupervised learning, logic, neural-symbolic integration, high-order neural connections

1 Introduction

The use of symbol processing and knowledge representation is fundamentally established in the field of Artificial Intelligence as a tool for modelling intelligent behaviour and thought. Specifically, logic-based systems have been used for years to model behaviour and cognitive processes. Nevertheless, it is believed that pure symbolic modelling is not enough to capture the adaptivity, sub-symbolic computation, robustness, and parallelism of the kind demonstrated by neural networks in the brain [20, 34].

ANNs are used today mainly for statistical pattern recognition, classification, and regression, and have been criticized for their weakness in representing, manipulating, and learning complex structured knowledge. citeFodor:1988,McCarthy:1988. For example, in areas such as vision and language processing, ANNs excel as classifiers and predictors. However, they are not very successful in recognizing and manipulating complex relationships among objects in a visual scene or in processing complex sentence structures governed by non-trivial semantic and syntactic constraints.

The main motivation of this paper is therefore to enable ANNs to have learnable combinatorial search and relational representation capabilities while using "practicing" to speed-up the search. Having such capabilities without sacrificing their statistical learning abilities will allow integration of the symbolic and statistical approaches for performing high-level cognitive tasks as motivated in previous research [3, 10, 11, 29, 34].

Several fundamental questions arise in the effort to represent and process complex knowledge using ANNS. First, there is a problem of representation and coding: How can objects and complex relational structures be encoded in the activation of neurons? This should be done without exponential explosion of network size, [7] and without losing accuracy as structural complexity grows [28]. Then, there is the problem of encoding learnable relational knowledge in connection weights. Also puzzling is the way to dynamically translate between the two forms of knowledge, i.e., retrieving structured relational knowledge from connection weights into activation encoding and in the opposite direction, storing new and revised active memories in the connection weights for later use. Finally, how can learning to search be done without the presence of a teacher who knows the correct answer for the search problem at hand?

The network architectures proposed, "Recurrent Network for Constraint Satisfaction" (CONSRNN) and "Symmetric Network for Constraint Satisfaction" (CONSyN), enable compact encoding of attributes and relationships in neural activation and storage of long-term relational constraints in connection weights. When provided with a problem to solve, such networks can execute iterative cycles of unit activation and learning, until a satisfying solution emerges on the output units. By unsupervised practicing on a training set composed of a few problem instances, the network can learn and improve its speed at solving unseen instances from a test set.

The two neural architectures are described, theoretically analysed, and empirically tested. In the implementations presented, flat networks with high-order multiplicative connections (sigma-pi units) are used with standard activation functions: sigmoidal in CONSRNN and binary-threshold in CONSyN.

The proposed ANNs facilitate learning by minimizing a loss function that is based on the domain constraints, which are provided as prior knowledge. When the loss is minimized, it means that the output activation values encode a solution that minimizes the constraint violation.

The paper is organized as follows: Section 2 illustrates an example of a simple block-world planning problem that will be learned and searched by the proposed ANNs. In section 3, we show how to represent the inputs and the solution output of the planning problem in activation of the network's visible units. Section 4 illustrates two different neural architectures for constraint satisfaction: CONSRNN, a network based on simple recurrence loop, and CONSyN, based on symmetric connections. Thereafter, sigma-pi units are reviewed, and two types of loss functions are introduced which measure how well a solution satisfies the constraints of the problem. The detailed activation-learning algorithm of CONSyN and its mathematical derivation are described in Section 5, while Section 6 describes the algorithm developed for CONSRNN. Section 7 describes the experimental framework and results. Section 8 discusses related disciplines, offers conclusions, and suggests directions for future research.

2 A Simple Planning Problem and Its Constraints

Although this article is about a general technique to learn logic constraints, for illustration, we have chosen the well-known problem of planning in block-world with a simple yet non-standard reduction to grounded FOL along the lines of "Planning as SAT" [18].

Consider a world of blocks, where some blocks happen to be on the floor, while other blocks are arranged "above" others. A planning agent has the task of arranging the blocks by a series of moves, so that a certain "goal" configuration of blocks is reached. The block-world changes with time, as the agent moves "cleared" blocks and puts them either on the floor or on top of other blocks. The result of the planning process is a "solution plan" which details a series of block-world configurations which are the result of corresponding moves. A solution plan therefore, is a series of configurations, that starts with the initial block configuration at time $t = 0$, ends in the goal configuration at time $t = K$, while each intermediate configuration at time $t = 1 \ldots K$ is a result of valid moves executed on the previous configuration at time

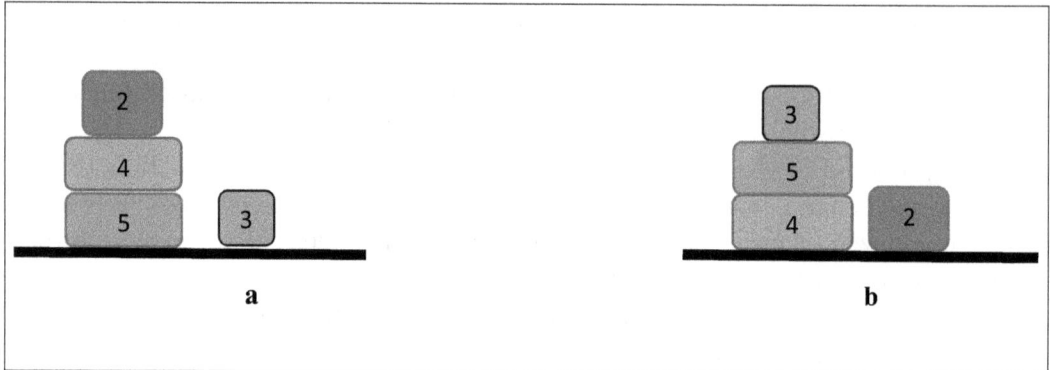

Figure 1: A planning problem instance: a) initial block configuration, b) goal configuration

$t - 1$. A valid move of block i to block j at time t, requires both blocks to be cleared at t; i.e there are no block above i, and no blocks above j at t unless it is the floor.

The relation $Above(i, j, t)$ specifies that block i is above block j at time step t. Note that although the Above relation completely specifies the world states and the plan, for ease of understanding and rule simplification, few other relations are added:

$Move(i, t)$ // block i moves at time t
$Cleared(i, t)$ // block i at time t is cleared (nothing on top)[1]
$Floor(i)$ // object i is a floor

A planning problem instance is specified by an initial configuration of blocks and a goal configuration. Figure 1a is an example of an initial configuration which includes four blocks. The goal configuration is to build a modified tower as in Figure 1b.

The plan that the agent must generate is a valid series of up to K block configurations in the "Above" relation, which by a series of corresponding valid "moves", gradually changes the initial configuration until the goal configuration is reached. Formally, a valid plan is specified by the "Above" relation if the logic constraints outlined in Table 1 hold. Though it is not the classic formulation of a planning problem, satisfying the constraints of table 1 is sufficient to produce classic block plans. For example, the $Move(i, t)$ relation is non-orthodox one as most other for-

---

[1]The *Clear* relation sometimes behaves rather unintuitively. For example, even when a block has no other block above it, it is not always cleared. Only when we move a block from block i to block j do these blocks need to be cleared just before the move. The cleared behavior can be "fixed" by adding more constraints, but this is unnecessary in our formalism.

1)	$(\forall \, i, j, k, t > 0) \; Above(i, k, t) \rightarrow$ $Above(i, k, t-1) \vee Move(i, t-1)$	If one object is above another at time t, it must be above that object in the previous step or have been moved in the previous step.
2)	$(\forall \, i \neq j, k, t) \; Above(i, k, t) \wedge$ $Above(j, k, t) \rightarrow Floor(k)$	If two objects are above an object then that one object is the floor.
3)	$(\forall \, i, t) \; \neg Above(i, i, t)$	No object can be above itself.
4)	$(\forall \, i, t, k \neq i) \; Above(i, j, t) \rightarrow$ $\neg Above(i, k, t)$	An object cannot be above more than one object.
5)	$(\forall \, i, j, t) \; Floor(j) \rightarrow \neg Above(j, i)$	The floor cannot be above any object
6)	$(\forall \, i, j, t) \; \neg Floor(j) \wedge$ $Above(i, j, t) \rightarrow \neg Cleared(j, t)$	If an object is above a second object that is not floor, then the second is not cleared.
7)	$(\forall \, i, t) \; Move(i, t < K) \rightarrow$ $Cleared(i, t) \wedge Cleared(i, t+1)$	If an object is moved at time t, then it is cleared at time t and at time $t+1$.
8)	$(\forall \, i, t) \; Move(i, t < K) \wedge$ $Above(i, j, t+1) \rightarrow Cleared(j, t)$	If an object was moved at time t to be above a 2nd block, then the 2nd was cleared at time t.
9)	$(\forall \, i, t, \exists j) \; Move(i, t) \rightarrow$ $Above(i, j, t)$	If an object was moved, then it was above something.

Table 1: Constraints of a simple block-world planning problem

mulations use a more explicit two-object "from-to" moves. The advantage of the shorter formulation is in a reduced network complexity; i.e., the less explicit move relation is more compact in terms of high-order connections, while the target object can be inferred from the above relation at $t+1$. Every valid plan must satisfy the above hard constraints and every solution to this constraint satisfaction problem can be proved to have the block plan properties; i.e. every configuration at time t is valid (e.g., a block is above only one other block) and a result of a valid move on the previous configuration. Nevertheless, we can add soft, non-monotonic rules [26]. Thus, for example, adding soft $(\forall i, t) \; (Cleared(i, t))$ clears all blocks unless there is something above them; and adding $(\forall i, t) \; \neg(Move(i, t))$ minimizes the number of moves, thus produces parsimonious plans. These soft rules are not mandatory for valid plan generation in our particular planning example; however, can save in rules and network connections, when for example, we wish to enforce frame axioms or parsimonious plans.

In a more general formulation, each constraint may be augmented by a positive number, called alternately "penalty," [26] "weight" in Markov Logic networks, [5] or "confidence" [33, 4]. A satisfying solution in our case is a solution that minimizes the sum of penalties of the violated constraints, when they are transformed into augmented CNF form as specified next. In our simple block-world planning domain, a penalty of 1000 is used to specify hard constraints, and a penalty of 1 is used for soft constraints.

In block-world planning of bounded length, the size of the maximal plan is restricted (K) and so is the maximal number of objects (N). These bounds are necessary in our architectures, as the full solution output should be encoded using a finite set of network units. These bounds are also used for reducing the above FOL expressions into propositional Conjunctive Normal Form (CNF), by replicating the constraints according to the indices specified. The reduction into propositional logic uses standard grounding. Thus for example, $(\forall i, t, \exists j)\ Move(j, t) \rightarrow Above(i, j, t)$ is translated into a CNF by replicating:

$for\ i = 1\ to\ N$
  $for\ t = 1\ to\ K$
    $AssertClause\ (\text{``}\neg Move(i, t)\ \lor\ Above(i, 1, t)\ Above(i, 2, t)\ \ldots \lor\ Above(i, N, t)\text{''})$

All planning instances including those of size (t) and object cardinality (n) smaller that the bounds $(n \leq N$ and $t \geq K)$, share the same propositions and the same hard and soft constraints specifying what a valid plan is, yet each planning instance is different in its initial and goal states. These initial and goal states are considered the inputs of the planning problem and can also be stated as simple conjunctive constraints. For example, the following conjunctions specify the initial configuration of Figure 1a:

$Floor(1) \land Above(2, 4, 0) \land Above(4, 5, 0) \land Above(5, 1, 0) \land Above(5, 1, 0) \land Above(3, 1, 0)$

where $t = 0$ is the first time step. The goal configuration at the final step of the plan, as in Figure 1b, is specified by:

$$Above(3, 5, K) \land Above(5, 4, K) \land Above(4, 1, K) \land Above(2, 1, K),$$

where $t = K$ is the final state.

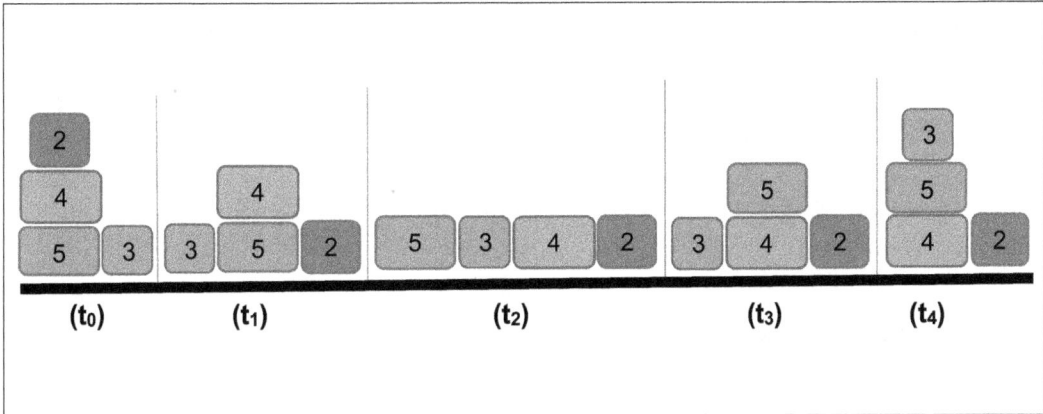

Figure 2: The solution plan to the planning instance of Figure 1

A valid solution for the planning instance of Figure 1 is illustrated in Figure 2. It consists of a series of valid block configurations in the "Above" relation which is the result of a series of valid moves in time:

$$Move(2,0), \ Move(4,1), \ Move(5,2), \ Move(3,3).$$

The block configurations at each time step is valid in the sense that it is consistent with all the hard constraints specified in Table 1. In Figure 2 at time 0, the initial block configuration holds, and only Blocks 2 and 3 are cleared. By moving Block 2 onto the floor, the configuration at $t = 1$ is created, where all blocks are cleared except block 5. At $t = 1$ Block 4 is moved onto the floor. At $t = 2$, the large green Block 5 is moved, and at $t = 3$ Block 3 is moved to create the desired goal configuration at time $t = 4$. No block moves further, so that the same block configuration remains static until the last time slot. The configuration of the last step therefore satisfies the goal constraints.

Randomly generated planning instances are used for training and testing. Each problem instance differs from others only in the initial and goal configurations; the rest of the constraints (enforcing plan validity) are shared among the different instances. A single ANN is created for solving any such planning instance of up to N blocks and up to K steps.

This architecture is independent of the specific constraints that will be learned and be stored as weights. The generated ANN will start solving a planning instance problem by first clamping the Boolean values of the Initial-Goal input onto the visible units of the network. In the following section, the structure of the visible units of such ANN is described.

3 Representing Relational Knowledge in Unit Activation

To attach attributes to objects and to represent relationships among them, a representation is needed. We base our representation on the binding mechanism published in Pinkas et al. (2013). Its dynamic capabilities are, however, not exploited in the simple planning example presented in this paper.

3.1 Forming Bindings to Represent Object to Object and Object to Property Relations

In the block-world planning problem of the figures above, the number of block objects is bound to $N = 5$ and the number of maximal planning steps to $K = 6$. To provide a "glue" that will allow the binding of objects with their properties, or with other objects collections of units called binders ($b1 \ldots b5$) are allocated to each of the objects that might participate in the plan. The network visible units consist of a pool of such multi-unit binders. In the general binding framework, [27] the binders from the pool are allocated dynamically to functions, constants and variables representing predicate logic formulae; however in our simple example, the object variables are grounded by attaching each object with a static binder. For example, binder $b3$ is a collection of units that was statically allocated block object 3. Binder $b1$ was allocated to the floor object. More generally, $b5$ could bind multiple properties (e.g., color, size) and k-tuple relation. In our experiments we have generated blocks with random colors and sizes, but no constraints were imposed on these properties.

In the block-world example, the binders are organized in 2D and 3D matrices of units called crossbars. Each unit has an activation function and depending on the network architecture, the activation value can be interpreted as either a Boolean value or a probability of being True. Although in the general binding framework, two crossbars are enough to represent any FOL formula, in our example, we use multiple crossbars, to encode several properties and relationships.

3.1.1 Using multi-dimensional Crossbars

The rows in a crossbar may also reference other binders, thus forming relationships among objects. For our planning example, the Above relationship should be 3D to capture the fact that the object configurations change over time. Therefore, the $Above(N, N, K)$ crossbar is used, where the object dimensions are bounded by N and the time dimension is bounded by K. Thus, $Above(i, j, t)$ means that object i is directly above object j at time t. Figure 3 illustrates the 3D Above crossbar

Above	t_0	t_1	t_2	t_3	t_4	t_5
b1						
b2	**4**	1	1	1	1	**1**
b3	**1**	1	1	1	5	**5**
b4	**5**	5	1	1	1	**1**
b5	**1**	1	1	4	4	**4**

Figure 3: The *Above* crossbar encodes the relationships between objects at each time step. The value j in cell $Above(i,k)$ means: $Above(i,j,k) = True$. The bold numbers represent clamped inputs.

encoding the changes in the Above relation over time in the solution plan of Figure 2, starting at the initial configuration and ending at the goal configuration. The input to the planner consists of the initial and goal configurations (in bold), which are clamped on the Above crossbar at t_0 and t_5, respectively. The configurations encoded at time steps t_1 to t_4 are generated by the planning agent and represent the solution found. In the example, at time t_1, the Above crossbar encodes the configuration after moving the block of Binder 2 onto the floor. At t_2, as a result of moving Block 5, all blocks are directly above the floor and all are cleared. The goal is reached at t4 after moving Block 3 above Block 5. At time t_4 no move is made, so the configuration remains static at t_5.

In Figure 4, the moves that should be executed by the plan are reflected within the 2D crossbar, $Move(N,K)$; the block referenced by Binder 2 is moved at t_0, creating the configuration encoded by $Above(,,t_1)$. The block referenced by binder 4 is moved at t_1, creating the configuration of $Above(,,t_2)$. Block 5 is moved at t_2 and finally, Block 3 is moved at t_3, generating the desired goal at $Above(,,t_5)$. The *Move* crossbar is not mandatory for representing the plan, as it can be deduced from the $Above(i,j,t)$ crossbar, yet its existence helps to specify shorter and more intuitive constraints. Note also that unlike classical SAT reductions of planning problem, $Move(i,t)$ does not specify where the object 1 is moved to.

Move	t_0	t_1	t_2	t_3	t_4
b1					
b2	**1**				
b3				1	
b4		1			
b5			1		

Figure 4: The $Move(N, K)$ relation encodes which objects are moved at each time.

Similarly, the 2D and 1D crossbars $Clear(N, K)$ and $Floor(N)$ are not mandatory but are also added for convenience. The $Clear$ crossbar describes which blocks are cleared at the various time steps, while the $Floor$ crossbar specifies which binder represents the floor object.

The visible units of the planning ANN are made of the set of crossbars: *Above, Move, Clear, and Floor.* These visible units consist of input units, which are clamped per planning instance according to the desired initial-goal configurations, and the output units, which encode the solution plan to be generated. Thus, the clamped inputs are the units of *Above, Clear* and *Floor* crossbars at t_0 (for the initial configuration) and at t_5 (for the goal configuration). The output units consist of crossbar units $Move$ and $Above(,, t)$, where $K > t > 0$.

Note that the *Above* relation, in this simple planning example, is a one-to-one relationship (per time instance): at time t, a block cannot be above two blocks and no two blocks can be above a single block. It is interesting to note that after these constraints are learned in the proposed symmetric architecture, the rows and columns of such crossbars turn to be "winner-takes-all" units, similar to the wiring suggested by Hopfield and Tank (1985).

3.2 Using Compact Distributed Representations

The encoding of the blocks used in this paper is one-hot encoding. If we allow distributed representations of the binders, binary relationships that are many-to-one may be represented compactly by using $N \log(N)$ units. Thus, the *Above* relation, which allows only one object to be above another single object, could be represented by $NK \log(N)$ units only. Similarly, an embedded representation for the blocks could also be used but is out of the scope of this paper.

Theoretically, this compact embedded representation reduces the search space. However, it may generate a biased input representation and may necessitate higher-order connections or deeper networks as the complexity of the constraints grows. Due to their complexity, the more compact representations were not tested experimentally.

4 ANN Architectures for Constraint Satisfaction

In this section, two different types of ANN architectures are sketched for solving constraint satisfaction problems. The first uses a symmetric matrix of weights and is based on the energy minimization paradigm, citeHopfield:1982,Peterson:1987 Boltzmann Machines, [1, 13] and Belief Networks [14]. The second architecture is a based on Recurrent Neural Networks (RNN) [35].

Both architectures have *almost* identical sets of visible units, where input values are clamped and where the output solution emerges as a satisfying solution. The rest of the units are hidden units, which, as shown later, can be traded with high-order connections. The visible units are directly mapped onto the problem's Boolean variables and should maximally satisfy the problem constraints. In the block-world example, the visible units are the units of the crossbars described earlier; the input units are those clamped by the initial and goal configurations, and the output units are those crossbars that at the end of the process encode the generated plan.

In order to solve a specific problem instance clamped on the input units, both network architectures iterate between activation calculation and learning phases. Activation calculation (with some stochasticity) computes the activation function of each of the units, while learning changes the connection weights, trying to minimize a loss function that relates to constraint violations. The iterations stop when a measure of the violation is "small enough," indicating that the process continues until all the hard constraints are satisfied and the number of violated soft constraints is less than a given parameter.

The idea is that practicing on solving training problem instances from the same domain will speed-up the solving of unseen instances; i.e., while solving several problem instances from the same domain, weights are learned that "better" enforce the domain constraints, thus decreasing the number of activate-learn iterations needed to solve test instances from the same domain.

Figures 5 and 6 illustrate the two architectures. Despite their similarities, the two architectures are very different in their dynamics and in their learning procedures.

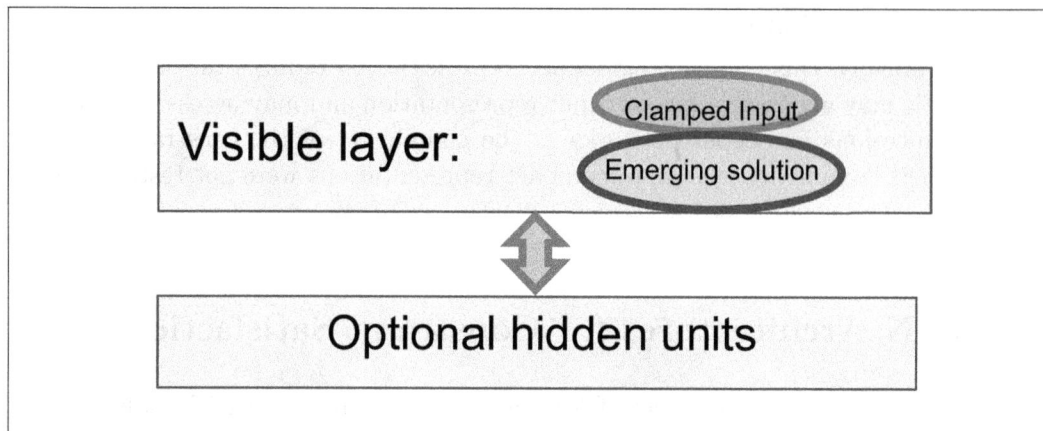

Figure 5: CONSyN architecture. The visible units are capable of encoding both the input and the desired solution. The hidden units are optional and may be traded with high-order connections.

4.1 Energy Minimization Symmetric Architecture

Figure 5 illustrates a symmetric ANN architecture for constraint satisfaction (CONSyN). The visible units consist of the problem crossbars and allow the encoding of both the input and the output. The input (init-goal) configuration is clamped onto the input part of the visible units while at a fixed point; the non-clamped (output) visible units get activations that ideally max-satisfy the constraints. Symmetric (possibly high-order) weighted connections and (optional) hidden units enforce the problem constraints on the visible units.

Whenever, the network reaches a unit-configuration that is a stable fixed, a

---

Algorithm 1 Activate-Learn iterations in CONSyN

Given a set of constraints, a problem instance input and a CONSyN network:

  a. Clamp the input (e.g. initial and goal configuration).
  b. Set random initial activation values to all non-clamped input units.
  c. Until a fixed point is reached, unsynchronously calculate activations of the non-clamped units.
  d. While violation loss is not "small" enough, do:
      i. CONSyN learning (Algorithm 3).
      ii. Until a fixed point is reached, calculate activations of the non-clamped units (Algorithm 4).

---

check is made to determine whether constraints are still violated. If that happens, a learning step is made, and the weights are updated in a way that increases the energy of the unwanted unit configuration.

Algorithm 1 describes in high-level the iterative procedure of activation calculation and learning until a "good enough" solution is found. When the activate-learn loop ends, the visible units have a violation measure that at least satisfy all the hard constraints.

The learning in the symmetric case (see next section) may be viewed as increasing the "importance" of the violated constraints. Such change in constraint importance is translated to Hebbian connection weight changes that in turn increase the energy of that violating state. Thus, learning is actually re-shaping of the energy function by incrementally lifting the energy of violating local minima. The result of the training is a network with an energy function that resembles a violation loss function.

4.2 Recurrent Network Architecture (CONSRNN)

The RNN in Figure 6 uses directed connections. The input layer consists of crossbar units capable of encoding both the clamped problem input (e.g. initial-goal configurations) and activation states copied from a feedback layer in previous recurrence. The output layer consists of crossbar units capable of encoding the generated plan. In our implementation, a simple feedback loop connects the output units to the nonclamped input units. However, the architecture is not limited to this simple form of

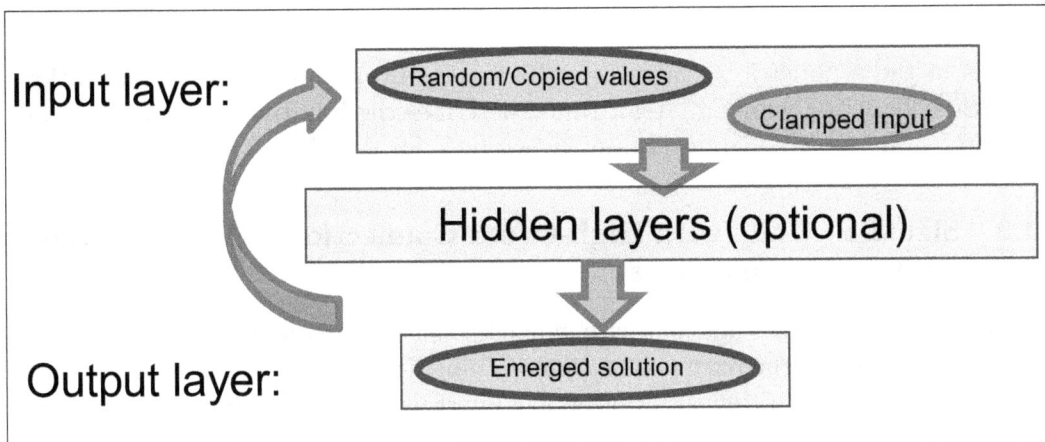

Figure 6: The CONSRNN architecture consists of a feedforward network with simple feedback loop. Activations from the feedback layer are copied onto the nonclamped input layer.

recurrence and future implementations could feed-back hidden units through time in a more standard way.

After clamping the inputs, the activation of the output layer is calculated by forward propagation. The output layer units together with the clamped input units fed into a loss function that measures constraint violations. If the violation of the output units is not "small enough," learning is done by truncated backpropagation through time, and the states of the output units are copied back into the non-clamped inputs (possibly with added noise) for another iteration of forward activation and learning. The clamped inputs stay clamped as in the previous iteration, and the process continues until a "good enough" solution emerges on the output units.

Algorithm 2 Activate-learn iterations in CONSRNN

Given a set of constraints, a problem instance input values and a CONSRNN network:

    a. Clamp the input values on the input units.
    b. Set random initial activation values to all non-clamped input units.
    c. Compute the activation of the output layer (by performing feed-forward calculation).
    d. While the violation loss is not "small enough," do:
        i. Back-propagate the gradient of a violation loss function through time.
        ii. Copy the feedback layer onto the (non-clamped) input layer.
        iii. Add noise to the non-clamped input units.
        iv. Compute the activation of the output layer.

As in the symmetric architecture, the search involves iterations of activation, violation check and learning. Algorithm 2 describes the iterative search for a "good enough" solution at a high level (see Algorithm 5 for detailed implementation).

4.3   Sigma-Pi Units with High-Order Connections and their Trade-off with Hidden Units

Unlike the classic units and pairwise synapses that are mainstream for many current neural models, the output of a sigma-pi unit is based on the sum of weighted contributions from multiplicative subsets of input values [6, 30, 36]. A sigma-pi unit is a generalization of a "classic" unit that uses high-order connections in addition to biases and pairwise connections. A k-order connection connects a set S of k units using a single weight. It can be a directed high-order connection (as in the feed-forward network of Figure 7) or a symmetric connection (Figure 8). Sigma-pi units

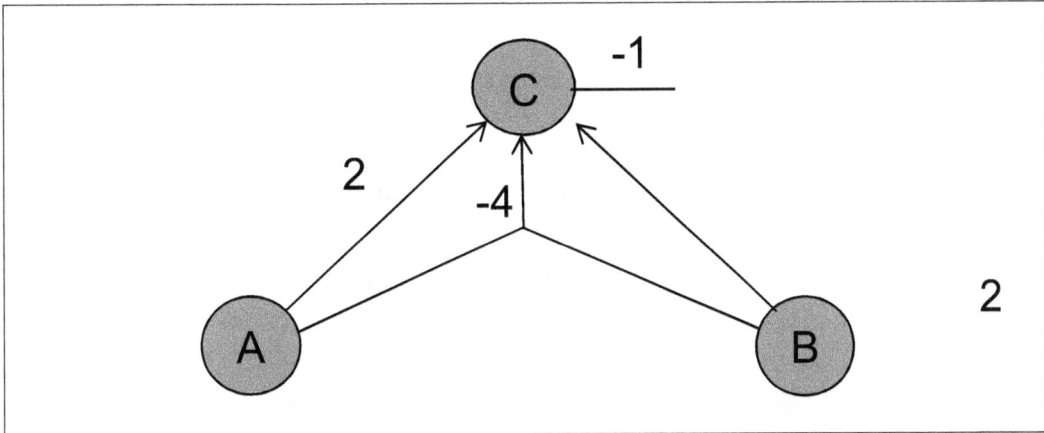

Figure 7: High-order feed-forward network for XOR: Unit B is a sigma-pi unit with a bias, 2 pairwise connections and a third-order connection.

calculate the weighted sum of products of the input units as in Equation 1.

$$z_i = \sum_{S_{j,i}} w_{s_j} \prod_{k \in S_{j,i} - \{i\}} X_k \tag{1}$$

To calculate the activation value for unit i, the weighted sum $(z_i,)$ of the connection products directing to i is computed. After calculating z_i, unit i calculates its activation value y_i using an activation function σ as in Equation 2.

$$y_i = \sigma(z_i) \tag{2}$$

Although a variety of activation functions could be used, the binary threshold activation function was used in our implementation of CONSyN, whereas sigmoidal activation was used in CONSRNN. In Figure 7, an example of a high-order feed-forward network for XOR is shown with a third-order directed connection $\{A, B\}_C$ that connects the product of Units A and B with Unit C and uses a single weight of -4. Unit C also has a bias of -1 (first-order connection) and 2 pair-wise connections $\{A\}_C$, $\{B\}_C$ (second-order) with weights of -2. Unit C therefore is a sigma-pi unit which calculates the weighted sum of input products: $Z_C = -4AB + 2A + 2B - 1$:

In the symmetric architecture, a k-order connection S_i is treated as if there were k such connections directing to each member of S_i, all with the same weight. Figure 8, illustrates an example of such high-order symmetric network. The network consists of four symmetric connections: A single third-order connection that connects all three units with weight of -2; two standard pairwise connections that connect units

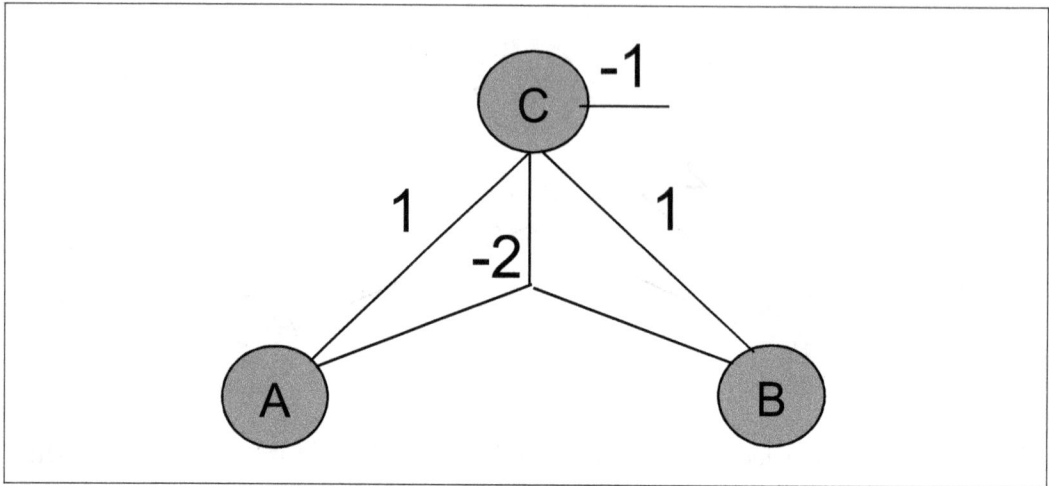

Figure 8: A symmetric third-order network with three sigma-pi units and a single third-order connection, searching to satisfy $C \to (A \text{ XOR } B)$.

A with B and B with C with weight of 1 and a negative bias for Unit B. Unit A computes therefore $Z_A = C - 2BC$, Unit B computes $Z_B = -2AC + C$, and Unit C computes $Z_C = -2AB + A + B - 1$. The network happens to search for a satisfying solution to $C \to (A \text{ XOR } B)$ as when $C = 1$, either A or B must be 1 exclusively.

Standard (pairwise) symmetric networks may be viewed as searching for a minimum of a quadratic energy function that is directly related to the weighted connections and may be written as a weighted sum of up to two variable products. Similarly, high-order symmetric networks minimize higher-order energy functions that sum the weighted products of the units in each of the connection subsets. Thus, the network shown in Figure 8 minimizes the third-order energy function: $E(A, B, C) = C - AC - BC + 2ABC$.

The connections are symmetric in the sense that a k-order connection is an undirected hyper-arc with a single weight, i.e. the tensor of weights is invariant under all permutations of its unit arguments. For example, in Figure 8, a single weight is attached to all connection permutations: $w_{\{A,B\}C} = w_{\{A,C\}B} = w_{\{B,A\}C} = w_{\{B,C\}A} = w_{\{C,A\}B} = w_{\{C,B\}A} = -2$.

High-order connections can be traded with Binary hidden units. Trivially so, in feed-forward networks, a directed k-order connection that adds $w \prod_{k \in S} X_k$ to the sum of weighted inputs z, can be replaced by a single hidden unit that performs an AND on the inputs using standard pairwise connections and a bias. Similarly, but not as trivially, symmetric k-order connections in the energy minimization paradigm can be replaced by at most k units with an equivalent energy function [24].

In the implementations of the two architectures described in this paper, high-order connections are used instead of hidden units. The lack of hidden units simplifies analysis of the network dynamics and reduces the search space hyper parameters. On the other hand, high-order connections may easily cause overfitting, and this may explain the degradation in network performance which was occasionally observed. Nevertheless, future implementations may choose to trade the high-order connections with deep layers of hidden units and thus will not need sigma-pi units.

4.3.1 Loss Functions

ANNs typically use loss functions to measure their functional error. The gradient of this loss function is used to incrementally update the connection weights of the network so that gradually, the error is minimized. Both suggested architectures use loss functions that measure the degree of constraint violation of activation values of the visible units. In both implementations, learning is done using Stochastic Gradient Descent (SGD).

To calculate the loss, we assume that the domain constraints are provided as a CNF input. Each literal in the CNF corresponds to a visible unit. The CNF violation is measured therefore with respect to the array of activation values of the output units. A clause is satisfied if at least one of its positive literals refers to a unit activation that is interpreted as "True" or at least one of its negative literals refers to a unit with "False" activation. The Violation Loss function (*Vloss*) in Equation 3 is a non-negative function that measures how "far" the activation values are from perfectly satisfying all the CNF clauses. A zero *Vloss* means that all clauses are perfectly satisfied, and a positive *Vloss* means that some clauses are violated, at least to a degree. More generally, in order to accommodate soft and hard constraints, as well as other probabilistic and non-monotonic logic systems, [26, 5] each of the clauses in the CNF is augmented by a positive penalty (α_c), which specifies the "strength" (or "confidence") of the constraint. The violation of an augmented CNF is the sum of the penalties of the violated clauses.

The activation values are in $[0, 1]$, thus enabling logics with probabilistic or belief interpretations.

$$Vloss(CNF, y) = \frac{1}{\sum_{c \in CNF} \alpha_c} \sum_{c \in CNF} \alpha_c \, ClauseLoss(c, y) \qquad (3)$$

4.4 Log-Satisfaction *Loss*

The function $ClauseLoss(c, y)$, described in the following sections, measures the degree of violation of a single clause c with respect to an activations array y. Note, that $ClauseLoss()$ has a probabilistic interpretation when y is assigned probabilities.

Equation 3 provides the Vloss function for measuring the violation of a given augmented CNF. From Equation 3, it follows that the gradient of the total $Vloss$ is the average of the gradients per clause. In the following, two different formulations of the $ClauseLoss$ function are provided: $Vprob$ and $LogSat$.

4.4.1 Vprob ClauseLoss

The $Vprob$ of a clause c with respect to activation array y is the product of the distances of the actual values y_l of the literals l in the clause from the desired values of that literals (Equation 4). The $Vprob$ $ClauseLoss$ function outputs a real number in the range [0,1], where 1 means perfect violation, and 0 means perfect satisfaction.

$$Vprob(c, y) = \prod_{l \in c} (1 - Lprob(l, c, y_l)) \tag{4}$$

The $Lprob$ function of Equation 5 measures the degree[0-1] to which the literal is satisfied. If probabilistic meaning is assigned to the unit activation, $Lprob$ represents the probability of satisfying the literal l in c.

$$Lprob(l, c, v) = \begin{cases} v & \text{if } l \text{ is a positive literal in } c \\ 1 - v & \text{else} \end{cases} \tag{5}$$

When all the activation values y corresponding to the clause variables are opposing their desired literal signs, the clause is violated and the $Vprob$ is exactly 1. If at least one of the variables has an activation that is exactly the desired value, the clause is satisfied and the product of the distances is 0. When the activation values are given probabilistic interpretation, the $Vprob$ function outputs a real number in $(0,1)$ which could be interpreted as the probability of the clause to be violated (under literal independence assumption). For example, consider a clause $c = (A \vee B \vee \neg C \vee \neg D)$ and activation array $y = [0.1, 0.2, 0.6, 0.7]$ corresponding to the truth probabilities of the units A,B,C,D, then

$$Vprob(c, y) = (1 - 0.1)(1 - 0.2)(0.6)(0.7) = 0.3024$$

The $Vprob$ $Vloss$ of a multi-clause CNF is the (weighted) average of the $Vprob$ $ClauseLosses$ per clause (Equations 3 and 4) and can be brought into a sum-of-weighted-products form. For example, assuming penalties of 1, the $Vprob$ Loss of the two-clause CNF, $(A \vee B \vee \neg C \vee \neg D) \wedge (\neg C \vee D)$, is the average of the two $Vprob$ $ClauseLoss$ functions: $\frac{1}{2}((1 - A)(1 - B)CD + C(1 - D)) = \frac{1}{2}((CD - ACD - BCD + ABCD) + (C - CD))$ which may be re-written as the weighted sum of products:

$0.5C - 0.5ACD - 0.5BCD + 0.5ABCD$. The *Vprob Vloss* is differentiable and its gradient with respect to the activation values is specified by Equation 6:

$$\frac{\partial V prob(c,y)}{\partial v} = \begin{cases} -\prod_{l \in c \wedge l \neq v}(1 - Lprob(l, c, y_l)) & \text{if } v \text{ is a positive literal in } c \\ \prod_{l \in c \wedge l \neq v}(1 - Lprob(l, c, y_l)) & \text{else} \end{cases} \quad (6)$$

For example, given $c = (A \vee B \vee \neg C \vee \neg D)$, the partial derivative of *Vprob* with respect to a positive literal A is:

$$\cdots = \frac{\partial(1-A)(1-B)CD}{\partial A} = -(1-B)CD = CD - BCD$$

Whereas the partial derivative with respect to a negative literal is:

$$\cdots = (1-A)(1-B)D = D - AD = BD + ABD$$

Notice that the partial derivative of the *Vprob* Loss is the average of the partial derivatives per clause and therefore also has the form of a sum of weighted products. A sigma-pi unit can therefore calculate the partial derivative of the *Vprob* loss using multiplicative connections. This ability of direct gradient computation by sigma-pi units enables the *Vprob* function to act as an energy function which is minimized by a symmetric architecture, rather than just as a loss function to guide learning.

Note also that when the chain rule is used (as in backpropagation) in conjunction with sigmoidal-like activation functions with input z, the *Vprob ClauseLoss* gradient is multiplied by $v(1-v)$, which is the derivative of the sigmoid. Equation 7 shows the partial derivative of the *Vprob ClauseLoss* with respect to z.

$$\frac{\partial V prob(c,y)}{\partial z} = \cdots$$

$$\frac{\partial V prob(c,y)}{\partial v}\frac{\partial v}{\partial z} = v(l-v)\begin{cases} -\prod_{l \in c \wedge l \neq v}(1 - Lprob(l, c, y)) & \text{if } l \text{ is a positive in } c \\ \prod_{l \in c \wedge l \neq v}(1 - Lprob(l, c, y)) & \text{else} \end{cases}$$

$$(7)$$

This may cause gradients to diminish when the units approach their extreme 0 or 1 values. Empirically this phenomenon was hardly observed in the implementation.

4.4.2   Log-Satisfaction loss

In another single-clause loss function, the log of the satisfaction degree is measured using Equation 8. The *LogSat ClauseLoss* returns therefore values between *0* (satisfaction) to *infinity* (violation) and may be interpreted as log likelihood under literal dependency assumptions.

$$logSat(c, y) = -log(max_{l \in c}\{Lprob(l, c, y)\}) \qquad (8)$$

Intuitively, when one of the literals is in close proximity to its desired value in the clause, the clause is "almost" satisfied. When this happens, the max returns a value near 1 and the "*log*" therefore is near 0. When the y values are near the opposite of their desired values, the "*max*" returns a value near 0, and the "*–log*" value approaches infinity. When probabilistic interpretation is taken, $logSat$ computes the probability of satisfying a clause assuming literal dependencies such as $p(A|B) = 1$ or $p(B|A)$. The partial derivative of $LogSat(c, y)$ with respect to a variable v in y is given in Equation 9 (ignoring non-differentiable points).

$$\frac{\partial logSat(c, y)}{\partial v} = \begin{cases} -\frac{1}{v} & \text{if } v \text{ is a positive literal in } c \text{ and has maximal } Lprob \\ \frac{1}{1-v} & \text{if } v \text{ is a negative literal in } c \text{ and has maximal } Lprob \\ 0 & \text{else } (Lprob \text{ is not maximal}) \end{cases} \qquad (9)$$

Equation 10 is a result of using the chain rule for calculating the $LogSat$ partial derivative with respect to z, assuming sigmoidal-like activations are used. Note that the result of multiplying Equation 9 by $v(1-v)$ has no diminishing terms problem.

$$\frac{\partial logSat(c, y)}{\partial z} = \cdots$$

$$\frac{\partial logSat(c, y)}{\partial v}\frac{\partial v}{\partial z} = v(1-v) \begin{cases} v - 1 & \text{if } y \text{ is positive and has maximal } Lprob \\ v & \text{if } y \text{ is negative and has maximal } Lprob \\ 0 & \text{else} \end{cases} \qquad (10)$$

When using SGD, the update step becomes proportional to $(1-v)$ for a positive literal and $(-v)$ for a negative one. This is intuitive, since it means that in a violated clause, the variable closest to its desired value should be incremented if it is a positive literal and decremented otherwise. The update step size should be proportional to the proximity $(1 - Lprob)$. For example, when $c = (A \vee B \vee \neg C \vee \neg D)$ and $y = [0.1, 0.2.0.6.0.7]$ the gradient of the $LogSat$ $ClauseLoss$ with respect to z's is the array $[0,0, 0.6,0]$, as C is the variable with maximal $Lprob$ to its desired value, and is a negative literal in c. To compute the $Vloss$ gradient when the CNF contains several clauses augmented by penalties, the gradients are computed for each clause and then averaged according to the penalties. Similar yet different log-likelihood loss which assumes literal independence, is possible but was not implemented.

5 Algorithms Implemented for Flat High-order CONSyN

In the CONSyN architecture, the *Vprob* Loss function is used for two purposes: In the first, it is used as a base for a high-order energy function that controls the dynamics of the network and is gradually modified by the learning process. In the second, the *Vprob* loss is used to guide the learning process, thus shaping the energy surface in order to find solutions faster. Although several symmetric ANN paradigms could be used (e.g., RBM, Belief Networks, MFT), a Hopfield-style network was implemented with no hidden units at all but with high-order connections instead.

When the units of the CONSyN are activated asynchronously, the network may be viewed as minimizing a high-order energy function using SGD. The units reverse their activations in a direction opposite to the direction of the energy gradient. Learning in this paradigm means that small changes are made to the energy surface when the network settles on a violating local minimum, so that the energy of violating states is "lifted" while the energy of non-violating states remains unchanged.

5.1 *Vprob Loss* as an energy function

The energy function that is being learned is specified in Equation 11. It is very similar to the Vprob loss function; however, there is no need to average and the weights (β) are learned and need not be identical to the preset penalties (α) of the loss function.

$$E_{CNF}(y) = \sum_{c \in CNF} \beta_c Vprob(c, y) = \sum_{c \in CNF} \beta_c \prod_{l \in c} (1 - Lprob(l, c, y)) \qquad (11)$$

The minimizing network consists of visible Hopfield-like sigma-pi units that correspond to the variables y of the energy, symmetric high-order connections that correspond to the product terms of the energy and connection weights which are the coefficients of the products terms. This network can be initialized automatically from the CNF with penalties that are preset ($\beta = \alpha$). For example, the network in Figure 9 is a result of pre-assigning a penalty of β_1, β_2 for the two clauses in the CNF $\beta_1(A \vee B \vee \neg C)$ and $\beta_2(\neg A \vee B)$. The energy minimized by that network is $E_{CNF}(A, B, C) = \beta_1(1 - A)(1 - B)C + \beta_2(A - AB) = \beta_1 C - \beta_1 BC - \beta_1 AC + \beta_1 ABC + \beta_2 A - \beta_2 AB$, while the network's weights (Figure 9) are the coefficients of the energy terms with opposite signs.

The units of Figure 9 compute (asynchronously) the energy gradient and reverse their activations away from the gradient direction, thus performing SGD down the energy.

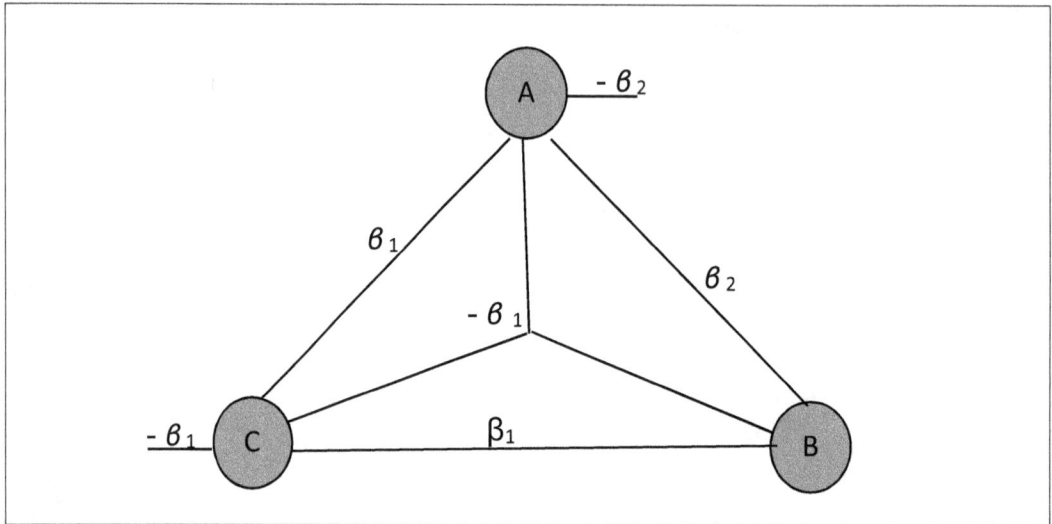

Figure 9: A high-order symmetric network that searches for a satisfying solution for $(A \vee B \vee \neg C) \wedge (\neg A \vee B)$ with corresponding learnable penalties $\beta 1$ and $\beta 2$.

When the β-penalties are all positive, Equation 11 is globally minimized in exactly the satisfying solutions of the CNF. When the CNF could not be satisfied, the global minima are those solutions with minimum violation.

In the experiments of the next section, an initial network has been generated in two varieties: The first is a "compiled" network generated using connections derived from the *Vprob* function with pre-set weights $\beta = \alpha$. The second is a "random" weight network where only the connection products are derived from the *Vprob,* while the weights are initialized randomly in [-1,1].

5.2 Searching for Satisfying Solutions in an Energy Minimization Network

As shown, by assigning positive penalties (β) in the energy of Equation 11, the generated CONSyN network searches for solutions with minimal violation, yet it may be plagued by local minima that only partially satisfy the CNF. The task of learning, therefore, is to adjust the connection weights to avoid local minima. The hope is that by using this kind of energy re-shaping, some local minima will disappear and the average time required to find a satisfying solution will be reduced. Although the learning process changes the energy surface, the β-penalties remain positives, and it is therefore guaranteed that the satisfying solutions of the CNF (if any) are equal to the global minima of the energy. The following is a specific implementation of Algorithm 1 for solving a problem instance using a flat high-order CONSyN network with Hopfield-like activations where SGD is used both during learning (weight gradient) and during activation (state gradient):

Algorithm 3 Implemented CONSyN constraint solver

Given:
- Augmented CNF - constraints augmented with penalties
- Inputs for clamping - a set of literals to be clamped for a particular problem instance
- CONSyN - with symmetric connections supporting the Vprob loss product terms

  Hyper-parameters:
  - *SelectedVClauses* - the number of violated clauses to learn from
  - *WeightBound* - a bound on the maximal connection weight allowed
  - *MaxSoft* - specifying the maximal number of soft constraint violations allowed

a. Clamp Inputs onto the input units.
b. Set random Binary activation values to all non-clamped units.
c. Calculate *Activation until Convergence* (Algorithm 4).
d. While (some hard constraints or more than MaxSoft soft constraints) are violated, do *Learn-Activate:*
    i. *CONSyN learning:*
        Randomly select violated clauses c (up to *SelectedVClauses* such clauses)
        For each selected c, do *SGD using Generalized Anti-Hebb* Learning:
            for each connection s containing the negative literals of c,
                - If the number of zero units in s is even, decrement weight w_s else increment w_s
                  (no need for learning rate as the weight change is scaled so that at least one clause unit flips value).
            - If $|w_s| > WeightBound$, downscale all weights by a factor of 0.01.
    ii. Calculate *Activation until Convergence.*

5.3 The Activation Calculation Process

In the activation step, the network sigma-pi units are asynchronously activated using the binary threshold function of Equation 12, until an energy minimum is obtained. Algorithm 4 describes the way activation fixed-point was implemented.

$$y = \begin{cases} 1 & \text{if } z > 0 \\ 0 & \text{if } z < 0 \\ Flip & \text{else} \end{cases} \quad (12)$$

Algorithm 4 Activation until Convergence calculation - Hopfield–like

---

Given:
- Symmetric network (CONSyN) and unit activations array
- *MaxRandmFlips:* the maximal number of consecutive random flips (hyper-parameter)

Repeat loop:
- a. *Must flips:* While there exists an unstable unit i; i.e. $(y_i = 1 \wedge z_i < 0)$ or $(y_i = 0 \wedge z_i > 0)$, randomly select such unit i and flip its value.
- b. *Random Flips:* if there exist unit i with $z_i = 0$, then randomly select such unit i and flip its value.
- c. If *MaxRandomFlips* consecutive random flips were done, exit loop.

The activation values y at the end of the loop are considered the local minimum of the energy function, despite the fact that the random walk in a plateau (Step b) may have been terminated too early (by Step c).

5.4 A Generalized Anti-Hebb Learning Rule for k-order Connections

In SGD, when clauses are violated, the weights of the network change in a direction opposite to the gradient of the *Vprob Loss*. This update step can be simplified into a Hebbian rule that is generalized to high-order connections (see Equation 16).

Intuitively, activations that violate constraints should be "unlearned," so that the energy associated with such activations is increased. This is done by weakening "supporting" connections and strengthening "unsupporting" connections in a process which reminds the "sleep" phase in "wake-sleep" algorithms [14].

In the following, we provide mathematical justification for this generalized anti-Hebb rule.

$$\Delta \beta_c = -\lambda \frac{\partial PropVloss(cnf, y)}{\partial \beta_c} = -\lambda \sum_v \alpha_c \frac{\partial ProP(c, y)}{\partial v} \frac{\partial v}{\partial z_v} \frac{\partial z_v}{\partial \beta_c} \quad (13)$$

The only learnable parameters in the energy of Equation 11 are the β_c's which "weigh" each clause. Therefore, when performing SGD down *Vprob Loss*, the update rule for β_c is obtained using the chain rule, where λ is the learning rate and α_c is the preset penalty of the augmented clause c (Equation 13). Notice however, that in the symmetric paradigm, each unit v, computes (zv) which is (minus) the partial derivative of the energy function as in Equation 14(a). Therefore, when computing the partial derivative of z with respect to β_c, Equation 14(b) is obtained. Combining Equations 13 and 14(b) results in Equation 15.

$$\text{(a)} \qquad\qquad\qquad\qquad\qquad\qquad\qquad \text{(b)}$$

$$z_v = -\sum_c \beta_c \frac{\partial Vprob(c,y)}{\partial v} \qquad\qquad \frac{\partial z_v}{\partial \beta_c} = -\frac{\partial Vprob(c,y)}{\partial v} \qquad (14)$$

$$\Delta\beta_c = \lambda \sum_v \alpha_c \beta_c \left(\frac{\partial Vprob(c,y)}{\partial v}\right)^2 \frac{\partial v}{\partial z} \qquad (15)$$

From Equation 15 and knowing that $\frac{\partial v}{\partial z} \geq 0$, it follows that the β-weight of a violated clause should always increase while executing SGD down the *Vprob Loss*. Although somewhat surprising, this result is quite intuitive; whenever the network falls into a local minimum that violates a clause, the learning process changes the connection weights in order to strengthen the β-weight for the violated constraints and therefore "lifts" the energy of the violating state.

A single clause may affect many connections, thus incrementing the penalty is not a local weight change as one would wish for a neural network learning. Luckily, from observing the *Vprob ClauseLoss* function, it is possible to deduce how each of the weights involved will change when increasing the β-penalty as follows.

Looking at the *Vprob Loss* of Equations 4 and 5, one can observe that the sum of products, that is the result of multiplying the proximities, consists of positive and negative product terms. Each relevant product term includes the negative literals of the clause, while its sign depends on the number of positive literals in the product. The sign is positive when the product involves an even number of positive literals and is negative when an odd number of positive literals are involved. For example, for Clause c1: $(A \vee B \vee \neg C)$, the *Vprob* product terms are: $+C - AC - BC + ABC$, and all include the negative literal C. Product terms with an even number of positive literals (such as ABC or C) are positive, and those with an odd number of positive literals (such as $-AC - BC$) are negative terms. When these energy terms are translated into connections, the signs are reversed for SGD. In the example, the weight of the third-order connection ABC, should decrease because it includes an even number of positive literals (A, B). Similarly, the weight of the pairwise connection BC increases because it consists of an odd number of positive literals (B).

Fortunately, the parity of the positive literals can be sensed from the activation of the units when the clause is violated. Upon violation of the clause, the unit values corresponding to the positive literals are 0s (representing Boolean *"false"*) while those corresponding to the negative literals are 1s (representing *"true"*). It is therefore only necessary to compute the parity of the *"false"* units. If the number of *"false"* units in a connection related to a violated clause is even, the weight should decrease, otherwise, the weight should increase (Algorithm 3-i). This parity rule may be put elegantly when the Boolean values of the units have bipolar representation (1 for *true*, -1 for *false*) as in Equation 16:

$$\Delta w_s = -\lambda \prod_{y_j \in S} y_j \tag{16}$$

where S is a symmetric connection of bi-polar units. When clause c is violated, a relevant connection includes all the negative literals which are all ones, whereas the positive literals are all -1. By multiplying the unit values, the parity of the number of -1s is calculated.

Intuitively, the anti-Hebb rule means that whenever a clause c is violated, its related connections are "unlearned" in the following way: if the number of false units in a connection is even, the connected units "excite" each other and therefore, to unlearn the connection should be weaken. Otherwise, when the number of false units is odd, the units "inhibit" each other. Thus, the connection should be strengthen.

To see that this is an extension of the familiar anti-Hebb rule, consider a pairwise connection (second-order). If the two units "fire" together (or "silent" together), then they should be "unwired," and the connection weight should weaken. If one of the neurons fires and the other is silent, there is an odd number of false units, and the "wiring" between them should be strengthened. Similarly, bias unlearning is modeled as a 1-order connection.

Note that in implementing Algorithm 3, the learning rate λ is determined automatically per each clause that is violated; i.e., for each violated clause, λ is set to be the minimal value that would cause at least one unit within the violated clause to flip its value.

6   Algorithms Implemented for High-order CONSRNN Architecture

As seen in Figure 6, the RNN architecture consists of a feed forward network for mapping the input layer into the output layer with a feedback loop. Although hidden units and deep architecture may be used, high-order connections are implemented instead, with no hidden layers at all. Thus, in our implementation, the output layer is then copied (with some noise added) back into the input layer. Algorithm 5 is the more detailed implementation of Algorithm 2, using flat sigma-pi output units, no hidden layers and truncated back propagation of only a single time step. The connections that are used in the implementations are those specified by the product terms of the *Vprob Loss* with the addition of full pairwise connectivity (input layer to output layer) and biases in the output layer.

Algorithm 5 Implemented CONSRNN

Given:
- Augmented CNF
- Inputs for clamping
- CONSRN network - with connections at least supporting the *Vprob Loss* product terms

   Hyper-parameters:
  - *Batch:* the number of learnings iterations before weight update is actually made
  - *NoiseLevel:* probability of randomizing a non-clamped input unit after each iteration
  - *NoImprove:* the number of non-improving iterations before reinitializing inputs
  - *MaxSoft:* the maximal number of soft constraints allowed in a solution

a. Clamp inputs.
b. Set random (0-1) initial activation values to the non-clamped input units.
c. *Feed-forward computation:* calculate activations for the output layer.
d. While (some hard constraints or more than MaxSoft soft constraints are violated), do (*Learn-Activate iteration*):
   i. *CONSRNN learning:* weight changes using the *noisy-δ-rule* (Algorithm 6),
   ii. Every *Batch* iterations, average the changes and update the weights.
   iii. If there is no improvement in violation for *NoImprove* iterations, restart by assigning random (0-1) values to the non-clamped input units
   iv. else, copy the output layer onto the non-clamped inputs, while randomizing (with probability *NoiseLevel*) the values of the non-clamped inputs.
   v. *Feed forward computation,*

Learning using backpropagation involves computing the gradient of the *Vloss* with respect to the weights, using the chain rule as in Equation 17, where S_v is a directed connection from a set S of input units to output unit v.

$$\Delta w_s = -\lambda \frac{\partial V loss(cnf, y)}{\partial w_{s,v}} = -\lambda \frac{\partial V loss(cnf, y)}{\partial z_v} \frac{\partial z_v}{\partial w_{s,v}} = \lambda \delta_v \frac{\partial z_v}{\partial w_{s,v}} \qquad (17)$$

The partial derivative of the $-V loss(cnf, y)$ function with respect to z_v, is the error (δ_v) per unit v. Equation 18 provides the δ-error for unit v as a weighted average of the errors per clause:

$$\delta_v = -\frac{\partial V loss(cnf, y)}{\partial z_v} = -\frac{1}{\sum_c \alpha_c} \sum_c \alpha_c \frac{\partial Clause Loss(c, y)}{\partial z_v} \qquad (18)$$

Algorithm 6 Noisy δ calculation

Given a CNF and an activation array y Given hyper-parameters:
  - *NoisyGradProb:* The probability of selecting a clause for random error calculation
  - *LearningRate:* a positive real
  - *Mode = VprobVloss/LogSatVloss:* The specific type of *Vloss* function to be used

  a. For each violated clause c in the CNF calculate the clause error:
      i. With *NoisyGradProb* probability, select a random unit v from clause c and compute its error to be $1 - v$ if v is positive in c and $-v$ otherwise.
      ii. Else $(1 - NoiseyGradProb$ probability), If $Mode = LogSatVloss$, select the variable with maximal *Lprob* among the clause variable. Calculate its -error using Equation 10;
        else, $(Mode = VprobVloss)$ for each unit v in the clause:
        calculate its -error using Equation 7.
  b. The total error δv is the weighted average of the clause-errors for v (Equation 18).

The update rule for a weight is derived the usual way (from Equation 17) and is provided in Equation 19: the delta rule for a high-order connection.

$$\Delta w_{s,v} = \lambda \delta_v \frac{\partial z_v}{\partial w_{s,v}} = \lambda \delta_v \prod_{y_j \in S} y_j \qquad (19)$$

7 Experimental Results

Both proposed architectures share the same experimental framework: A weighted CNF was generated from the block planning constraint specifications (Table 1) allowing up to six blocks of up to three colors and three sizes and a maximum of seven step plans. The CNF uses 385 variables and 5,832 clauses. Hard clauses were augmented with $\alpha = 1000$ penalties, and soft clauses with $\alpha = 1$ penalties. Following the CNF generation, a network (either CONSRNN or CONSyN) was generated with visible units corresponding to the CNF variables and connections derived from the $Vprob(CNF, y)$ $Vloss$ function. The generated network was given a training set of planning instances to solve. Each training instance was solved by the network, and the weights learned during the solving were carried to the next training instance. Every 10 training instances, the network was tested by solving a set of 50 test instances.

The weights of the CONSRNN were randomly initialized, while in the CONSyN architecture, a comparison was made between the performance of random weight initialization and the performance of compiled weights initialization ($\beta = \alpha$). Algorithm 3 (for CONSyN) and Algorithm 5 (for CONSRNN) were used for both training and testing instances. However, during testing, the learned weights were not transferred from one instance to the next.

7.1 Generation of Train and Test Planning Instances

Random block-world instances were generated (450 instances) in three different levels of difficulty: 150 3-block problems (*easy*), 150 4-block problems (*medium*), and 150 5-block problems (*difficult*). Each problem instance included a random initial block arrangement, and a random goal arrangement. For each level of difficulty, 100 instances were used as a training set, and 50 different instances were used as a test set. A set of 50 3-block instances was used as validation to select some of the hyper-parameters.

As a proxy for performance, for each test instance, the number of steps required to obtain a satisfying solution was measured. In the CONSyN architecture, the number of unit flips was measured, whereas in the CONSRNN case, the number of activate-learn iterations was measured.

In each experiment, the test set was tried prior to training (point 0 in all graphs) and then again every 10 training instances. Each experiment was repeated 10 times, each time with a different (randomized) order of the training instances. The performance measured was averaged across the 50 test instances and then across 10 different experiments (with different randomization).

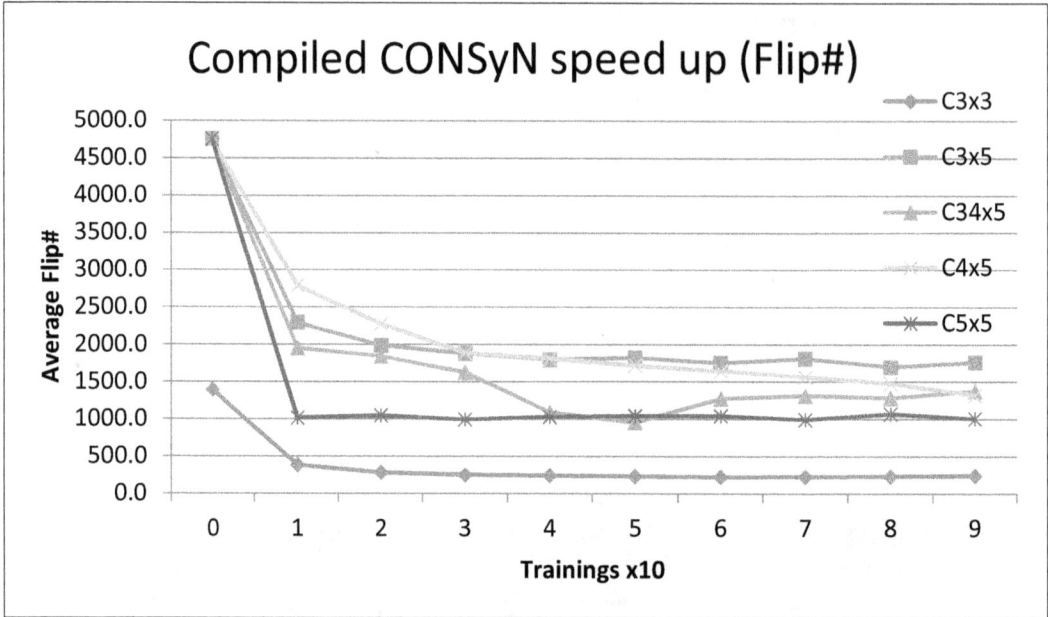

Figure 10: Performance (in flips) starting with compiled weights

7.2 Results for the CONSyN Architecture

The experiments of the symmetric network were conducted using Algorithm 3 and 4 running on a symmetric network generated with connections that correspond to the *Vprob Loss* terms of the CNF. Performance of a test instance was measured by the number of unit flips that were made until a solution was found. Figure 10 shows the average performance of a compiled CONSyN trained using training sets of various difficulty levels. Hyper-parameters used: $SelectedVClauses = 1$, $WeightBound = 200,000$, and $MaxSoft = 100$.

The C3 × 3 graph shows the test performance prior and during a training session of 100 3-block instances. The untrained (compiled) network starts at an average of about 1,400 flips per instance. However, after 10 training instances, the network gains speed (average flips is 385) and after just 20 training instances, the average number of flips for solving a 3-block test problem is stabilized around 250 flips.

The rest of the graphs in Figure 10 are tested on more difficult (5-block) instances. The C5 × 5, C4 × 5, and C3 × 5 graphs show the result of a 5-block test performance while practicing on 5, 4, and 3 blocks respectively. Not surprisingly, training on difficult (5-block) problems gives the best result. Surprisingly, the network is capable of generalizing from practicing on easy problems when tested on the

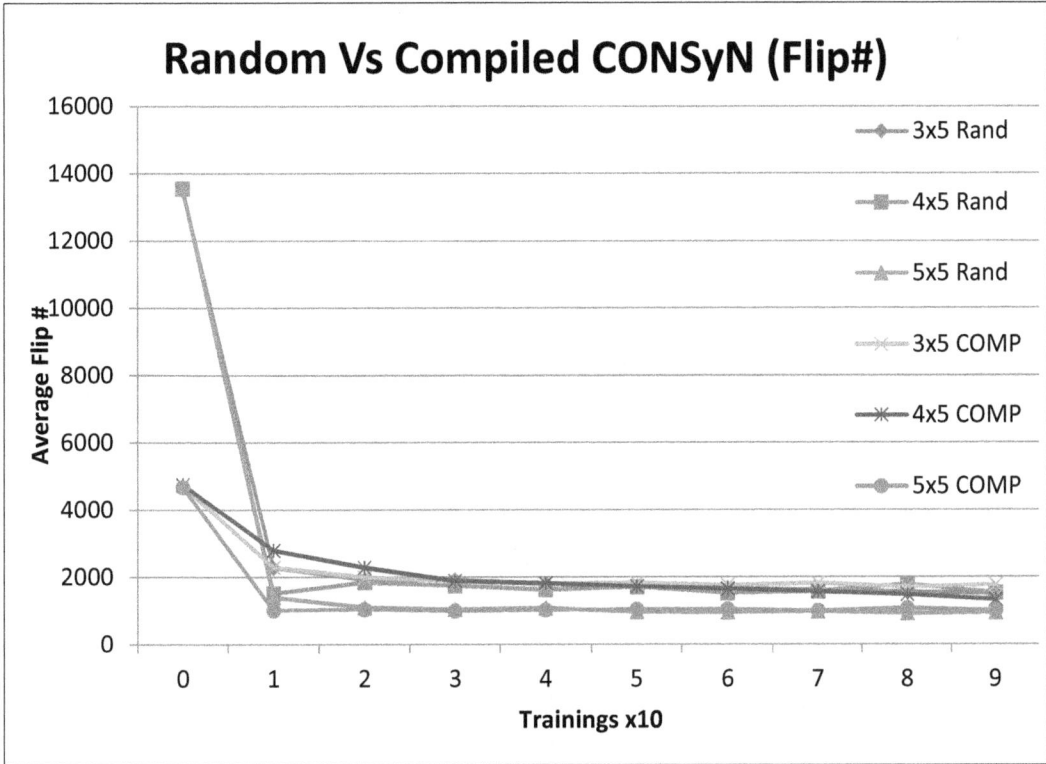

Figure 11: Comparison of randomly initiated networks with compiled networks (5X5)

more difficult 5-block training set. Training using four blocks speeds-up better than training on three blocks. However, training first on 20 easy (3-block) instances and then on the more difficult four blocks achieved the performance of 5-block training after about 60 trainings (see $C34 \times 5$ graph). However, when continued with 4-block trainings, the test performance degraded and approached the speed of the $C4 \times 5$ graph.

In Figure 11, the performance of compiled networks on 5-block test problems is compared with randomly initiated networks (Rand). The random networks are extremely slow just prior to training, but after few training instances, performance is accelerated and approaches that of the compiled network.

In Figure 12, the performance of a compiled CONSyN network of 5-block training and 5-block testing was measured using the average number of "activation-learn" iterations instead of the number of flips. The shapes of the graphs measuring iterations seem to resemble those measuring flips. Prior to the training, it took an

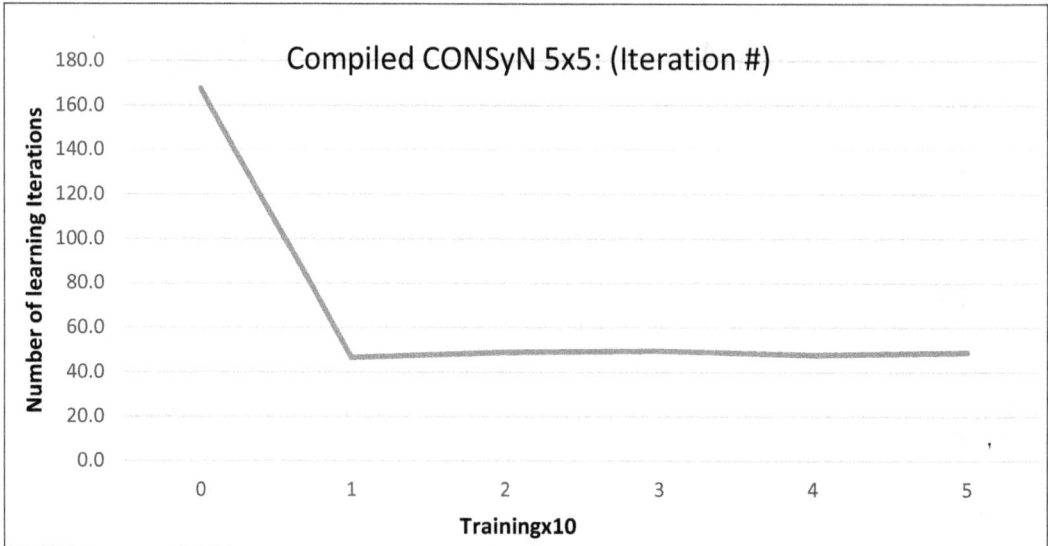

Figure 12: Compiled CONSyN network (5×5) measured using the number of iterations

average of 170 iterations to solve a 5-block test instance and after just 20 training instances, the number of trainings needed dropped to about 50 iterations on average. This may signal that the number of interfering local minima drops.

7.3   Results for the CONSRNN Architecture

The experiments of the CONSRNN were conducted using Algorithm 5 running on an RNN with high-order connections and no hidden units. The connections generated included all pairwise connections between the input and output layers, as well as all higher-order *directed* connections that corresponded to all permutations of the *Vprob Loss* product terms. For example, if ABC is a term in the *Vprob Loss*, then three directed 3-order connections are generated corresponding to the permutations of this particular term: A, B_C; A, C_B; B, C_A. The weights were initialized with uniform random real values in $(-1, 1)$.

The performance of the CONSRNN was tested using the same methodology. However, the number of "activate-learn" iterations was measured instead of the number of flips. As in the CONSyN case, each experiment was replicated 10 times with different randomizations, and the performance measured was averaged across the test instances.

Figure 13 shows the average number of iterations needed for solving 3-block

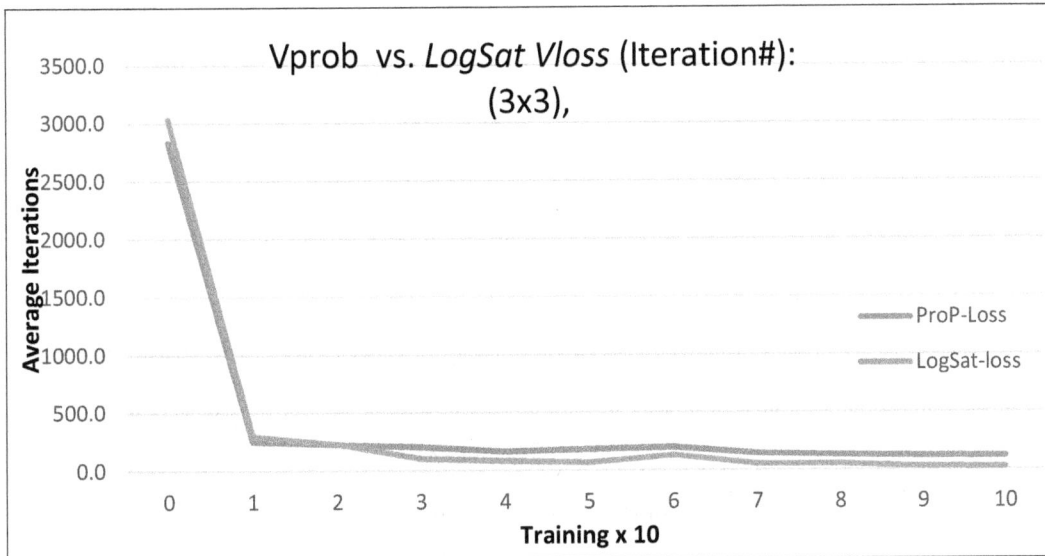

Figure 13: Comparing performance of *Vprob* vs. *LogSat Loss* (3×3)

instances while training on 3-block instances. The graph compares the result of using the *Vprob Loss* vs the *LogSat Loss* and shows similar speedups. In these experiments hyper parameters were selected using the 3-block validation set: $NoiseLevel = 0.15$, $NoisyGradProb = 0.06$, $LearningRate = 0.06$. Similarly, Figure 14 shows the performance speed-up when testing and training 4-block problems, while Figure 15 shows testing and training of 5-block problems.

Empirically, on this set of experiments, the *Vprob Loss* provided better performance than the *logSat loss* and was less prone to over-fitting.

In Figure 16, the effect of various noisy gradient probabilities on generalization after just 10 training instances are shown on 3-block training and testing.

Noisy gradient probabilities between 0.07 and 0.23 were tried; when the gradient noise level was too small (< 0.12), practicing was ineffective and even worsened the performance (0.07). Best performance and generalization were observed with noisy gradient probabilities of 0.16-0.18. Higher noise levels (e.g., $NoisyGradProb = 0.23$) seemed to be able to speed up performance but also led to overall inferior solving abilities.

In Figure 17, the effect of various noisy gradient probabilities is shown on a 5-block test set, while training used 4-block instances. Significant speed-up was observed immediately after the first 10 training instances. However, the test performance deteriorated when training continued as if the network was over-learning. With further training though, this deterioration was corrected in some experiments.

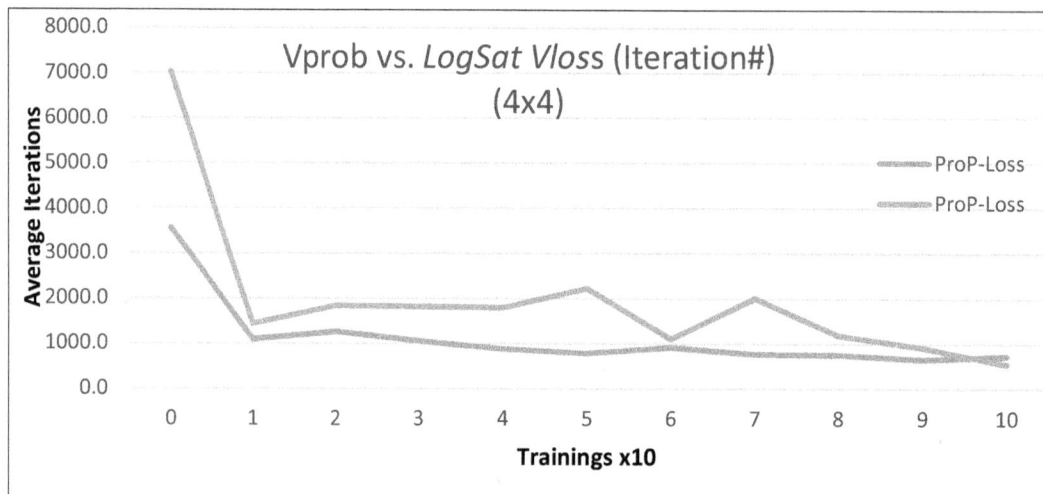

Figure 14: Comparing performance of *Vprob* vs. *LogSat Loss* (4×4)

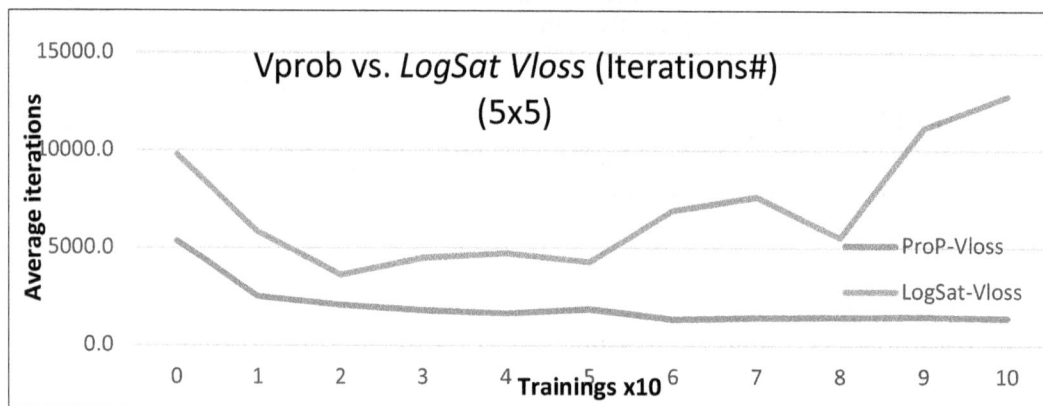

Figure 15: Comparing performance of *Vprob* vs. *LogSat Loss* (5×5)

8 Related work

Planning problems have been stated as logic deductions since the early days of artificial intelligence [8]. The idea of reducing planning problems into satisfiability (SAT) was first introduced by Kautz and Selman (1992) [18]. SAT solvers are used today for a variety of applications from planning to program verification and are considered among the fastest solutions for applied np-hard problems [12]. Some sophisticated SAT solvers learn on the fly while solving a specific problem instance, yet to the best of the authors' knowledge, SAT solvers currently do not carry learned knowledge from one problem instance to another. The ANNs proposed in this paper

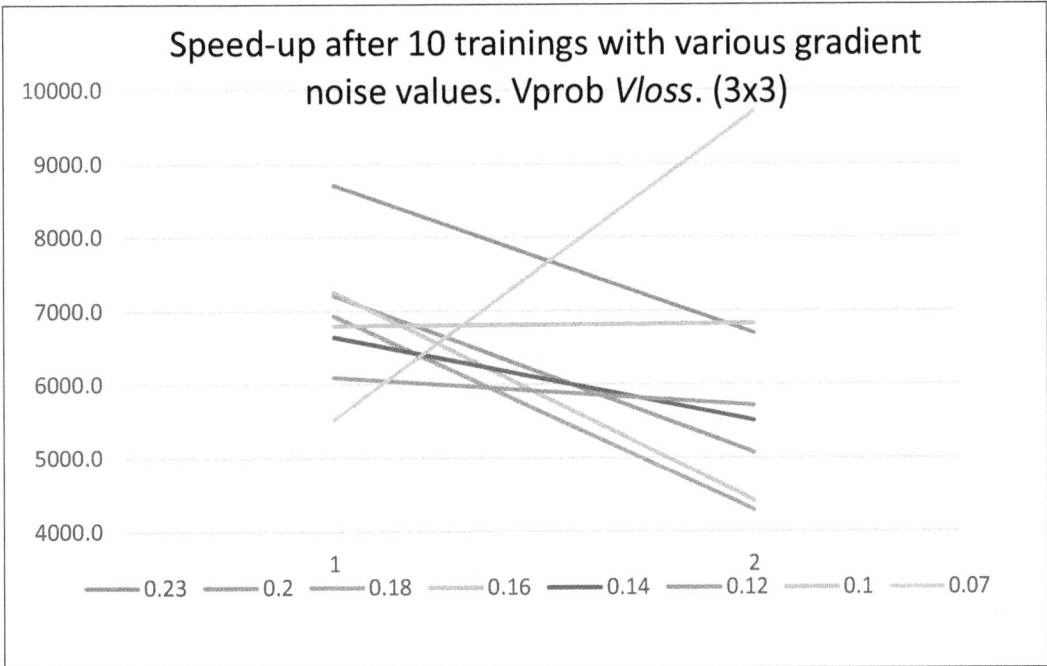

Figure 16: Speed-up of *Vprob Loss* (3×3) with various noisy gradient probabilities. Performance was measured before and after practicing 10 training instances.

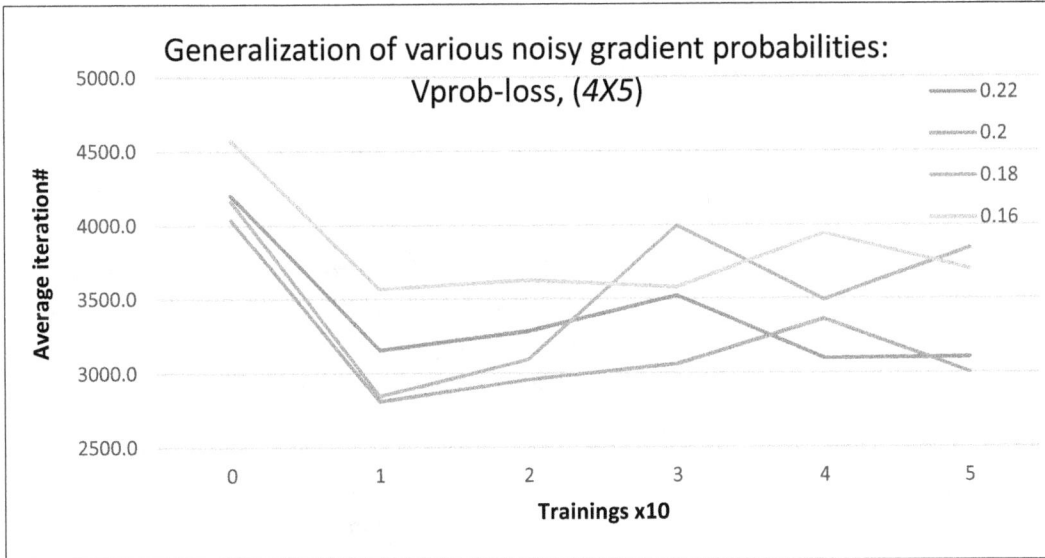

Figure 17: Effect of various noisy gradient probabilities on generalization (5-block testing, 4-blocks training)

may be viewed as SAT solvers that learn a particular application domain and adapt to it by transferring the learned weight from one problem instance to the next. It is thus interesting to note that the iterative loop in both CONSRNN and CONSyN have some resemblance to the basic loop of local search techniques such as GSAT and WALKSAT [31, 19]. In some sense, both architectures perform a strategy that may remind the local search greedy heuristics used by such SAT solvers. Unlike these solvers, though, the heuristic is learned by the ANNs and can be integrated easily with statistical learning.

In our example of planning, we did not use the well-established "planning as SAT" specification [18]. The specification used (Table 1) was chosen because it is consistent with the previously published binding mechanism and because it generates a more compact representation in terms of the order of connections and their number [27].

In the area of neuro-symbolic integration, [10, 11] emphasis has been given on how symbolic knowledge can be extracted from and injected into ANNs. The CON-SyN architecture is similar in its nature to a more general architecture designed to find proofs in bounded FOL ANNs [25, 21]. Adding learning ability on top of such systems is a key contribution of this article. In Tran and Garcez (2018), [33] augmented conjunctive rules were compiled prior to learning, and this was empirically shown to speed-up learning. However, the rules are limited in their structure and must be hierarchical in their nature. In the architectures proposed, augmented logic expressions with no structure limitations could be injected by either pre-compiling them (as in compiled CONSRNN) or by learning them using a *Vloss* function. Thus, prior knowledge can be injected before, during, and even after doing statistical learning.

8.1 Discussion and Summary

Two ANN architectures have been introduced that learn to search in a combinatorial search space, which is restricted by a set of logic constraints expressed using bounded FOL. The iterative activate-learn process employed by both architectures may be viewed as a parallel constraint satisfaction, using an adaptive learnable heuristic. The constraints are activated by the inputs) and are learned upon failing to satisfy them. After several iterations of activation and learning, a solution emerges on the output units, which satisfies the hard constraints with no more than a pre-specified number of soft constraint violations. The learned constraints may be viewed as long-term knowledge for solving a problem, which is stored in connection weights and retrieved when there is a need to solve an instance of the problem.

The learning process is guided by a loss function that measures the degree of constraint violation of the visible unit activations. Unsupervised learning is used to speed-up network performance by "practicing" on training problem instances. Significant speed-up is observed in the simple block planning domain after practicing on just 10 training problems when testing on unseen problem instances. Speedup is also observed when training is made on "easy" problems, while testing is done on more "difficult" problems.

Two loss functions have been proposed for measuring the degree of constraint violation. Empirically, a slight advantage has been observed for the *Vprob Loss* over the *LogSat Loss* in the CONSRNN architecture in some experiments. Nevertheless, we still consider both functions as valid implementation options. In the CONSyN architecture, only the *Vprob Loss* was implemented, as it naturally fits the energy minimization paradigm. The *Vprob Loss* was used both to pre-compile the network and to reshape the energy function while learning. The compiled network performs much faster than a randomly initialized network prior to training. However, the random network catches-up rapidly and obtains the performance of the compiled network after just a few training instances. In some experiments, it has been observed that the network slows-down when learning continues past a certain point. This may suggest "over-fitting" and therefore, the future use of strategies such as early-termination, drop-out, or regularization.

8.2 Some Notes and Insights

During the experimentation, we noticed that the networks did not find the obvious two step strategy of disassembling all blocks on the floor and then using them to construct the goal. A variety of different plans were generated as solutions with no noticeable bias towards the floor. Here are some informal insights on this behaviour:

- As far as problem specifications, there is almost no difference between a floor and a cleared block. Therefore, there was no bias towards using the floor for disassembling.
- The disassemble strategy is not always the most efficient one. As there is a bound K on the length of the plan, there are solvable instances which cannot be solved using this strategy.
- As both network architectures are designed to learn just greedy heuristics, the stochastic components of the network are important parts of its ability to solve the problem. As a by-product, solving the same problem again and again may result in a variety of solutions.

In theory, every search problem that can be reduced to SAT can be implemented in the proposed architectures, yet learning by practicing may only be useful if the problem instances (train and test sets) share many of the constraints, so that the learned constraints could be transferred from one instance to another.

8.3 Future Directions

There are plenty of technical variations yet to be explored in both architectures. Only a few of them have actually been implemented thus far. In the implementations presented here, no hidden units were used in a simple vanilla RNN. However, since high-order connections could be traded with layers of hidden, deep networks can be used instead of (or in conjunction with) high-order connections [2]. Thus, adding hidden units and extending the depth of the unfolding may enable the network to discover features related to the search history. The ample existing research in deep learning could be useful. For example, LSTMs could be used to learn strategies spanning longer iterations [15]. Convolution nets and inception nets may use variable size filters to detect certain local clause features within the multi-dimensional crossbars [32].

The computational mechanism demonstrated in this paper may be adapted for a variety of other logic or symbol-driven applications, such as verification and language processing. However, it is important to note that, in their current state, our simulated ANN implementations are not a match for the efficient state of the art current SAT solvers. The potential of this ANN approach may come from its domain adaptively and in its integration with statistical learning. Integrating supervised learning with prior knowledge could be executed before, after, or in conjunction with constraint learning. This could be done by combining familiar loss functions with *Vloss* functions.

For illustration, consider a network trained with a loss function that integrates *Vloss* with cross-entropy classification using a weighted average of the two functions. Such an approach would allow the mixing of symbolic constraint processing and statistical classification at the same time with varying confidence levels on the data and on the constraints. Thus, for a visual scene to be analyzed, objects and relationships could be classified while at the same time, the classification results should satisfy certain domain and physical world constraints. Both the data and the constraints may be augmented with varying confidence levels. The error, computed by the gradient of such integrated loss function will encapsulate both the statistical error and the prior knowledge violation error.

References

[1] D.H. Ackley, G.E. Hinton, and T.J. Sejnowski. A Learning Algorithm for Boltzmann Machines (PDF). *Cognitive Science*, 9(1):147–169, 1985.

[2] Y. Bengio. Learning Deep Architectures for AI. *Foundations and Trends in Machine Learning*, 2(1):1–127, 2009.

[3] Tarek R Besold, Artur D'avila Garcez, Sebastian Bader, Howard Bowman, Pedro Domingos, Pascal Hitzler, Kai-Uwe Kuehnberger, Luis C Lamb, Daniel Lowd, Priscila Machado Vieira Lima, Leo de Penning, Gadi Pinkas, Hoifung Poon, and Gerson Zaverucha. Neural-Symbolic learning and reasoning: A survey and interpretation. 11 2017.

[4] H. L. H. De Penning, A. S. D'Avila Garcez, Luís C. Lamb, and John-Jules C. Meyer. A neural-symbolic cognitive agent for online learning and reasoning. In *Proceedings of the Twenty-Second International Joint Conference on Artificial Intelligence - Volume Volume Two*, IJCAI'11, pages 1653–1658. AAAI Press, 2011.

[5] Pedro Domingos. Markov logic: A unifying language for knowledge and information management. In *Proceedings of the 17th ACM Conference on Information and Knowledge Management*, CIKM '08, pages 519–519, New York, NY, USA, 2008. ACM.

[6] J. A. Feldman and D. H. Ballard. Connectionist models and their properties. *Cognitive Science*, 6(3):205–254, 1982.

[7] Jerome Feldman. The neural binding problem(s). *Cognitive Neurodynamics*, 7(1):1–11, 2013.

[8] R.E. Fikes and N.J. Nilsson. STRIPS: A new approach to the application of theorem proving to problem solving. *Artificial Intelligence*, 2(3-4):189–208, 367, 1971.

[9] J.A. Fodor and Z.W. Phylyshyn. Connectionism and cognitive architecture: A critical analysis., 1988.

[10] A.d. Garcez, L. Lamb, and D. Gabbay. *Neural-Symbolic Cognitive Reasoning. Ser. Cognitive Technologies*. Springer, 2009.

[11] B. Hammer and P. Hitzler, editors. *Perspectives of Neural-Symbolic Integration*. Springer, 2007.

[12] Frank van Harmelen, Vladimir Lifschitz, and Bruce Porter. *Handbook of Know-ledge Representation*. Elsevier Science, San Diego, 2007.

[13] G.E. Hinton and R.R. Salakhutdinov. Reducing the dimensionality of data with neural networks. *Science*, 313(5786):504–507, 2006.

[14] Geoffrey E Hinton, Simon Osindero, and Yee-Whye Teh. A fast learning algorithm for deep belief nets. *Neural Computation*, 18(7):1527–1554, 2006.

[15] S. Hochreiter and J. Schmidhuber. Long short-term memory. *Neural Computation*, 9(8):1735–1780, 1997.

[16] J. J. Hopfield. Neural networks and physical systems with emergent collective computational abilities. *Proceedings of the National Academy of Sciences of the USA*, 79(8):2554–2558, 1982.

[17] J.J. Hopfield and D.W. Tank. "Neural" computation of decisions in optimization problems. *Biological Cybernetics*, 52(3):141–152, 1985.

[18] H. Kautz and B. Selman. Planning as satisfiability. In *Proceedings of the 10th European Conference on Artificial Intelligence*. Wiley, 8 1992.

[19] H. Kautz and B. Selman. Pushing the envelope: planning, propositional logic, and stochastic search. In *Proceedings of the Thirteenth National Conference on Artificial Intelligence*, pages 1194–1201, 1996.

[20] R. Kowalski. *Computational Logic and Human Thinking: How to be Artificially Intelligent*. Cambridge University Press, Cambridge, 2011.

[21] P. M. V. Lima. *Resolution-Based Inference on Artificial Neural Networks*. PhD thesis, Department of Computing, Imperial College London, U.K., 2000.

[22] J. McCarthy. Epistemological challenges for connectionism. *Behavioral and Brain Sciences*, 11(01):44, 1988.

[23] C. Peterson and J.R. Anderson. A Mean Field Theory Learning Algorithm for Neural Networks. *Complex Systems*, 1:995–1019, 1987.

[24] G. Pinkas. Symmetric neural networks and logic satisfiability. *Neural Computation*, 3(2):282–291, 1991.

[25] G. Pinkas. Constructing syntactic proofs in symmetric networks. In R.P. Lippmann. J.E. Moody, S.J. Hanson, editor, *Advances in Neural Information Processing Systems*, volume 4, pages 217–224, 1992.

[26] G. Pinkas. Reasoning, non-monotonicity and learning in connectionist networks that capture propositional knowledge. *Artificial Intelligence Journal*, 77(2):203–247, 1995.

[27] G. Pinkas, P. Lima, and S. Cohen. Representing, binding, retrieving and unifying relational knowledge using pools of neural binders. *Biologically Inspired Cognitive Architectures*, 6:87–95, 2013.

[28] T.A. Plate. Holographic reduced representations. *IEEE Transactions on Neural Networks*, 6(3):623–641, 1995.

[29] Alexei V Samsonovich. On a roadmap for the BICA challenge. *Biologically Inspired Cognitive Architectures*, 1:100–107, 7 2012.

[30] T J Sejnowski. Higher-order Boltzmann Machines. In *AIP Conference Proceedings 151 on Neural Networks for Computing*, pages 398–403, Woodbury, NY, USA, 1987. American Institute of Physics Inc.

[31] B. Selman, H. Kautz, and B. Cohen. Noise Strategies for Local Search. In *Proc. AAAI*, pages 337–343, 1994.

[32] Christian Szegedy, Wei Liu, Yangqing Jia, Pierre Sermanet, Scott Reed, Dragomir Anguelov, Dumitru Erhan, Vincent Vanhoucke, and Andrew Rabinovich. Going deeper with convolutions. 9 2014.

[33] S. N. Tran and A. S. d'Avila Garcez. Deep logic networks: Inserting and extracting knowledge from deep belief networks. *IEEE Transactions on Neural Networks and Learning Systems*, 29(2):246–258, 2 2018.

[34] Leslie G Valiant. Knowledge Infusion: In Pursuit of Robustness in Artificial Intel-

ligence. In Ramesh Hariharan, Madhavan Mukund, and V Vinay, editors, *IARCS Annual Conference on Foundations of Software Technology and Theoretical Computer Science*, volume 2 of *Leibniz International Proceedings in Informatics (LIPIcs)*, pages 415–422, Dagstuhl, Germany, 2008. Schloss Dagstuhl–Leibniz-Zentrum fuer Informatik.

[35] P. Werbos. Backpropagation through time: what it does and how to do it. *Proceedings of the IEEE*, 78(10):1550–1560, 1990.

[36] Ming Zhang. *Artificial Higher Order Neural Networks for Modeling and Simulation.* IGI Global, 2012.

Received 7 June 2018

Learning Representation of Relational Dynamics with Delays and Refining with Prior Knowledge

Yin Jun Phua
Tokyo Insitute of Technology, Japan
`phua@il.c.titech.ac.jp`

Tony Ribeiro
Laboratoire des Sciences du Numérique de Nantes, France
`tony.ribeiro@ls2n.fr`

Katsumi Inoue
Tokyo Institute of Technology, Japan
National Institute of Informatics, Japan
`inoue@il.c.titech.ac.jp`
`inoue@nii.ac.jp`

Abstract

Real world data are often noisy and fuzzy. Most symbolic methods cannot deal with such data, which often limit their application to real world data. Symbolic methods also have a problem of being unable to generalize beyond the training dataset. As such, it is very difficult to apply such methods in fields like biology where data is rare and fuzzy. On the other hand, neural networks are robust to fuzzy data and can generalize very well. In this paper, we propose a method utilizing neural networks, that could learn the representation of the relational dynamics of systems with delays. By training a recurrent neural network that learns the general pattern of several n-variable systems, it is able to produce a useful representation of another unseen n-variable system. This representation can then be used to predict the next state of the system when given k previous states. We show that our method is robust to fuzzy data, and also show that it can generalize to data that are not in the training data. Further, we also show that by giving prior knowledge to the model, we are able to improve the prediction accuracy of the model.

Vol. 6 No. 4 2019
Journal of Applied Logics — IfCoLog Journal of Logics and their Applications

1 Introduction

Learning the relational dynamics of a system has many applications. For example, in multi-agent systems where learning other agents' behavior without direct access to their internal state can be crucial for decision making [6]. In system biology, learning the interaction between genes can greatly help in the creation of drugs to treat diseases [12].

Having an understanding of the dynamics of a system allows us to produce predictions of the system's behavior. Being able to produce predictions means that we can weigh between different options and evaluate their outcome from a given state without taking any action. In this way, learning about the dynamics of a system can aid in planning [8].

In most real world systems, we do not have direct access to the rules that govern the systems. What we do have, however, is the observation of the systems' state at a certain time step, or a series of observations if we look long enough. Therefore, the problem is to learn the dynamics of systems purely from the observations that we are able to obtain.

Several learning algorithms have been proposed. One such algorithm is the Learning from Interpretation Transition (LFIT) algorithm [5]. The LFIT algorithm, when given a series of state transitions, outputs a normal logic program (NLP) that describes and realizes the given transitions.

The LFIT algorithm has largely been implemented in two different methods, the symbolic method [10] and the neural network (NN) method [14]. The symbolic method utilizes logical operations to learn and induce logic programs. One of the main flaws of the symbolic method, is that the logic program it produces cannot be generalized to transitions that are not present in the training dataset. It can sometimes be very expensive or impossible to obtain new observations. Thus, this flaw limits the possible application of this method. Another problem is that any noise present in the dataset, is reflected directly in the logic program. If the dataset is noisy and ambiguous, the algorithm will not be able to learn the logic program that is actually intended. Most observations obtained from the real world are fraught with unwanted noises, thus this problem also limits the application of the symbolic method.

On the other hand, the NN method trains a NN that models the target system, and then utilizes analytical methods to extract the logic program. The NN method described in [14] solves the flaw of being unable to generalize to unseen transitions in the symbolic method. However, the method in [14] does not deal with systems that contain delays. Most biological systems contain delays, therefore the logic program learned by [14] does not reflect the true relation of the system. Also, NN methods have no way of incorporating background knowledge. In cases where having background knowledge is useful for obtaining the intended model, this method does not provide an easy method to do that.

In this paper, we propose a method that utilizes NNs to learn the representation of the relational dynamics of systems with delays. This method aims to solve the problem of symbolic method not being able to generalize to unseen data, while also handling delays in the systems. It has the added advantage of being robust to noisy and ambiguous data, and generalizing well from very few training data. The representation produced by this method can also be used for prediction in a very scalable way. In particular, a prediction can be obtained by performing a matrix multiplication of the produced representation and an encoded vector of up to k previous states.

The rest of the paper is organized as follows. We cover some of the prior researches in Section 2, following by introducing the logical background required in Section 3. Then we present our representation learning approach in Section 4. We pursue by presenting an experimental evaluation demonstrating the validity of an approach in Section 5 before concluding the paper in Section 6.

2 Related Work

In this section, we survey several recent related work involving both neural and symbolic methods of learning relational dynamics. There are work in this area that learns probabilistic models. Here, we focus on deterministic approaches and specifically on methods that output logic programs.

2.1 Standard LFIT

One way of implementing the LFIT algorithm is by relying on a purely logical method. In [5], such an algorithm is introduced. It constructs an NLP by doing bottom-up generalization for all the positive examples provided in the input state transition. A method for learning smaller, much more interpretable NLP was introduced in [10]. In [11], an algorithm that learns delayed influences, that is cause/effect relationship that may be dependent on the previous k time steps, is introduced. Another recent development in the prolongation of the logical approach to LFIT is the introduction of an algorithm which deals with continuous values [13].

This class of algorithms that utilizes logical methods, are proven to be *complete* and *sound*, however a huge disadvantage with these methods is that the resulting NLP is only representative of the observations that have been fed to the algorithm thus far. Any observations that did not appear in the input, will be predicted as either to be always true or always false depending on the algorithm used.

2.2 NN-LFIT

To deal with the shortcomings stated in the previous paragraph, an algorithm that utilizes NN was proposed [14]. This method starts by training a feed-forward NN to model the system that is being observed. The NN, when fully trained, should predict the next state of the system when provided with the current state observation. Then, there is a pruning phase where weak connections inside the NN are removed in a manner that doesn't affect the prediction accuracy. After the pruning phase, the algorithm extracts rules from the network based on the remaining connections within the NN. To do so, a truth table is constructed for each variable. The truth table contains variables only based on observing the connections from the outputs to the inputs of the trained and pruned NN. A simplified rule is then constructed from each truth table. In [14], it is shown that despite reducing the amount of training data, the resulting NLP is still surprisingly accurate and representative of the observed system. However, this approach does not deal with systems that have inherent delays.

2.3 RNN-LFkT

In order to deal with systems that contains delay, and also to use NN in order to avoid the shortcomings of the logical method, a method that utilizes recurrent neural network (RNN) was proposed in [9]. This method trains a model that learns to extract general features from various systems, and provides a logic program representation that is distinct for separate systems but are the same for distinct state transitions from the same system. One shortcoming of this method is the inability to provide background knowledge to refine the logic program representation produced. Our work described in this paper is based on this method, attempting to overcome the shortcoming of being unable to provide background knowledge.

2.4 Other NN-based Approaches

There are also several other approaches attempting to tie NNs with logic programming [3, 4]. In [3], the authors propose a method to extract logical rules from trained NNs. The method proposed deals directly with the NN model, and thus imposes some restrictions on the NN architecture. In particular, it was not made to handle delayed influences in the system. In [4], a method for constructing NNs from logic program is proposed, along with a method for constructing RNNs. However this approach requires background knowledge, or a certain level of knowledge about the observed system (such as an initial NLP to improve on) before being applicable.

In [7], the authors proposed a method for constructing models of dynamical systems using RNNs. However, this approach suffers from its important need of training data, which increases exponentially as the number of variables grow. This is a well-known computational problem called the curse of dimensionality [2].

3 Background

The main goal of LFIT is to learn an NLP describing the dynamics of the observed system. An NLP is a set of rules of the form

$$A \leftarrow A_1 \wedge A_2 \wedge \cdots \wedge A_m \wedge \neg A_{m+1} \wedge \cdots \wedge \neg A_n \tag{1}$$

where A and A_i are propositional atoms, $n \geq m \geq 0$. $\neg$ and $\wedge$ are the symbols for logical negation and conjunction. For any rule R of the form (1), the atom A is called the head of R and is denoted as $h(R)$. The conjunction to the right of $\leftarrow$ is called the body of R. We represent the set of literals in the body of R as $b(R) = \{A_1, \ldots, A_m, \neg A_{m+1}, \ldots, \neg A_n\}$. The positive literals in the body is denoted as $b^+(R) = \{A_1, \ldots, A_m\}$, while the negative literals in the body is denoted as $b^-(R) = \{A_{m+1}, \ldots, A_n\}$. The set of all propositional atoms that appear in a particular Boolean system is denoted as the Herbrand base $\mathcal{B}$.

An Herbrand interpretation I is a subset of $\mathcal{B}$. For a logic program P and an Herbrand interpretation I, the immediate consequence operator (or T_P operator) is the mapping $T_P : 2^{\mathcal{B}} \mapsto 2^{\mathcal{B}}$:

$$T_P(I) = \{h(R) \mid R \in P, b^+(R) \subseteq I, b^-(R) \cap I = \emptyset\}. \tag{2}$$

Given a set of Herbrand interpretations E and $\{T_P(I) \mid I \in E\}$, the LFIT algorithm outputs a logic program P which completely represents the dynamics of E.

To describe the dynamics of a system changing with respect to time, we can use time as an argument. In particular, we will consider the state of an atom A at time t as $A(t)$. Thus, we can rewrite the form (1) into a dynamic rule as follows:

$$A(t+1) \leftarrow A_1(t) \wedge A_2(t) \wedge \cdots \wedge A_m(t) \wedge \neg A_{m+1}(t) \wedge \cdots \wedge \neg A_n(t). \tag{3}$$

Rule (3) means that, when $A_1, \ldots, A_m$ is true at time t and $A_{m+1}, \ldots, A_n$ is false at time t, then the head A will be true at time $t+1$. By describing all rules in an NLP in the form (3), we can simulate the state transition of a dynamical system with the T_P operator.

In [11], the LFIT framework has been extended to deal with systems that contain delays. Systems with delays can be considered as a Markov(k) system, in which the next state depends on up to k previous states. To represent rules in such systems, we can use the form (3) and substitute the argument t in the body as $t - j$ where $0 \leq j \leq k$.

4 Proposed Method

In this section, we describe our method of learning the representation of relational dynamics. In our model, we don't learn the representation directly. An optimal representation is obtained by training the model on a separate task, mainly the regression of the next state. Intuitively, by learning the regression on the next state, the NNs can focus on learning the abstract features of the particular system.

On a high level, we want to abstract the information represented in the logical space, into a more compact linear space. By abstracting and thus avoiding the need to deal with information in the lower level logical space, we are able to handle the fuzziness and ambiguity of the data. Our model thus can be thought of as first encoding information in the logical space, into the representation space. We then perform the T_P operator, that will give us the next state, in the representation space. Once we obtain the next state in the representation space, we can then map it back into the logical space.

Given an Herbrand interpretation I, we can define a vector $\vec{v}$ that represents the interpretation by setting each element in the vector v_i as 1 if $A_i \in I$ and 0 otherwise. Since we are dealing with a delayed system setting, we can consider a matrix $M_k = [\vec{v}_t \cdots \vec{v}_{t-k}]$ that contains up to k states.

We want to calculate the next state of the system given past k' states. Note that $k' \geq k$ can be greater than or equal to the true maximum delay of the Markov(k) system. For abbreviation purposes, we will refer to this k' as k, and will explicitly state the true delay of the system when required. When only looking at k-step state transitions, they can be produced by two totally different systems. In order to make distinction, we require some *a priori* knowledge about the system in order to distinguish between them. In the LFIT framework, this can be done by supplying an initial NLP. The equivalent in our model is done by supplying an initial representation $L_0 \in \mathbb{R}^{d \times d}$.

Thus, our model can be defined as calculating the following

$$\vec{v}_{t+1} = f_{\text{decode}}(f_{\text{representation}}(M'_k, L_0) \times f_{\text{encode}}(M_k)^\top) \qquad (4)$$

where M'_k can be equal or different from M_k, L_0 is an initial representation, $f_{\text{decode}} : \mathbb{R}^d \mapsto \{0,1\}^{|\mathcal{B}|}$, $f_{\text{representation}} : \{0,1\}^{|\mathcal{B}| \times k} \times \mathbb{R}^{d \times d} \mapsto \mathbb{R}^{d \times d}$, $f_{\text{encode}} : \{0,1\}^{|\mathcal{B}| \times k} \mapsto$

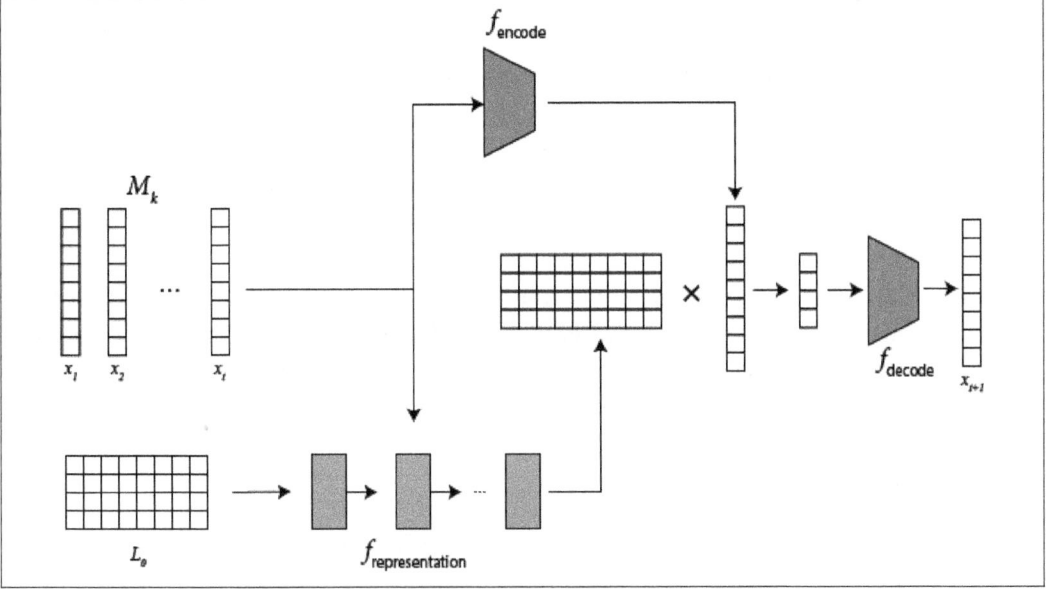

Figure 1: A visualization of our model

$\mathbb{R}^d$, d is the dimension of the learned representation. All of f_{decode}, $f_{\text{representation}}$ and f_{encode} are differentiable, $\times$ represents the matrix-vector multiplication, and therefore this entire model is end-to-end differentiable.

Learning is performed by minimizing the mean squared loss defined as follows

$$\text{loss} = \frac{1}{|\mathcal{B}|} \sum_{i=1}^{|\mathcal{B}|} (v_{t+1,i} - y_i)^2 \tag{5}$$

where y_i is the true label.

The f_{decode} function can be thought of as converting state vectors in the representation space, back into the logical space. The f_{encode} function does the opposite of converting state vectors in the logical space into the representation space.

The $f_{\text{representation}}$ function takes 2 parameters. Namely, they are the past k states that provides the abstract feature to extract from, and an initial representation that is the *a priori* knowledge. With these 2 parameters, the function outputs a matrix that is the representation of the dynamics of the system.

Implementation-wise, all of the above functions can be implemented with any non-linear function with parameters that can be trained by performing gradient descent. In our implementation, the f_{decode} and the f_{encode} functions are implemented with a multi-layer perceptron. The $f_{\text{representation}}$ function is implemented with LSTM (Long-Short Term Memory).

5 Experiments

We first verify that our model is capable of performing the designated task of predicting the next state. Next, we perform several experiments that exposes mainly the weaknesses of symbolic methods, namely inability to handle noisy and fuzzy data, and also being unable to generalize from the training dataset.

We implemented our model in Tensorflow [1]. Next, we performed the following experiments: first, the model is given discrete, error-free data, then we test the model on data that is fuzzy and ambiguous, lastly we test the model and data that are erroneous. As far as we are aware, there are no other works that perform such tests, particularly in delayed setting. Therefore we did not perform any comparisons but rather show the precise results in the experiments.

5.1 Hyperparameters

We did not perform too much tuning on the hyperparameters, mainly because the initial hyperparameters that we chose (rather randomly), worked well for most of the experiments. Here are the hyperparameters chosen when performing the following experiments:

- Training epochs: 3

- Batch size: 64

- Gradient descent optimizer: Adam, learning rate is set to 0.01

- All parameters are initialized randomly by the default initialzer in Tensorflow r1.5.

- No dropout or other regularization method is used.

- The MLP for f_{encode} consists of 1 hidden layer with 10 hidden nodes.

- The MLP for f_{decode} has 2 hidden layers, each with 10 hidden nodes.

- The implementation for $f_{\text{representation}}$ function is a 4-layer LSTM, each with 32 units.

- The representation dimension d is set to 4.

- The output threshold is set to 0.5.

In our experience, we did not find that tuning the above hyperparameters affect the experimental results by a huge margin.

5.2   Experimental Method

To train the model, we used a randomly generated dataset. Recall that our model aims to extract features of *any* given system. In particular, the constraint of the system is such that the system can be described by an NLP. Therefore, it is to our advantage that we can randomly generate a huge amount of data based on that constraint, and then train the model to extract common features of systems with such constraint.

Data generation works by first randomly generating an NLP, then generating state transitions that start from all possible initial states. However, if we purely randomly generate an NLP, we might not be able to generate *good quality* NLPs that allows the model to learn. Therefore, we limit ourselves to only generating NLPs with several properties. First, the body of each rules in the NLP should not be too long. We perform a random exponential cut-off for the length of the body, so there are cases where the body is long, but that should not be too often. Second, the state transitions generated from the NLP needs to have high variation. State transitions that are in the middle of an attractor are excluded from the training data. If there are too few state transitions from the NLP, then the NLP is also excluded from the training data.

During the experiment performed below, we generated 30,000 different NLPs. Then from each NLP, we obtained a maximum of 500 samples, each sample containing state transition for 10 timesteps. This gave us 150,000 samples to train the model.

Based on our experience, generating training data that has sufficient variance hugely affects the performance of the model. When we tried purely randomly generating NLPs and their corresponding states, we got transitions that are zero for most of the time or are constantly at the same state, and the model was unable to learn any useful features.

5.3   LFIT Benchmarks

We tested our model on 4 LFIT benchmarks. These are the same benchmarks that were also used in [5] and [10]. Note that these benchmarks do not contain any delays, but the results here will serve as a reference on how well our model is working on standard tasks.

We generated a series of 10 transitions from all possible initial states for each benchmark. The model is then asked to predict the next state based on these 10 transitions. We first ran all the benchmarks without providing any background knowledge, that is supplying the $\mathbf{0}$ matrix for L_0 in equation 4. Next, for predictions

Method	Mammalian (10)	Fission (10)	Budding (12)	Arabidopsis (16)
Without L_0	0.224	0.063	0.218	0.146
With L_0	0.184	0.062	0.199	0.128
Fuzzy data (25%)	0.209	0.062	0.206	0.134
Fuzzy data (50%)	0.243	0.072	0.215	0.238
Error (10%)	0.223	0.081	0.198	0.157
Error (20%)	0.249	0.101	0.211	0.201
Error (30%)	0.287	0.129	0.227	0.230
Error (40%)	0.334	0.182	0.250	0.251
Error (50%)	0.379	0.250	0.288	0.263
Error (60%)	0.418	0.324	0.318	0.274
Error (70%)	0.435	0.363	0.341	0.280
Error (80%)	0.466	0.406	0.353	0.300
Error (90%)	0.490	0.469	0.355	0.317

Table 1: The MAE of the prediction performed by our model on 4 separate benchmarks.

that are wrong by 1 or more variables, we obtained new L matrix by calculating $f_{\text{representation}}$ by supplying different state transitions than M_k, and then using that as L_0 for M_k to calculate the predictions again. The results are shown in table 1. We calculated the mean absolute error (MAE) as the metric for accuracy. An MAE of 0.2 for a 10 variable benchmark means that the model predicted 2 of the 10 variables wrong. As can be seen from the results, we were able to improve the predictions when providing L_0, except in the case of the fission benchmark where the model was already doing very well.

Next, we added fuzziness to the data by mapping each element in the state vector $\{0, 1\} \mapsto [0, 1]$. When a particular variable is 1, the value is fuzzed into a range of $[0.5, 1]$, and when it is 0 it is mapped into the range $[0, 0.5]$. The results of this is indicated by the row fuzzy data (50%). For fuzzy data (25%), we mapped 1 to the range of $[0.75, 1]$ and 0 to $[0, 0.25]$. In this experiment, whenever the model made wrong predictions, we provided L_0 that is calculated based on discrete error-free data. This is done under the assumption that in a real world scenario, the prior knowledge provided is usually considered as a fact. Therefore we did not do any fuzziness test on L_0.

We then tested the model's ability to handle erroneous data. For 10%, we flipped 10% of the variable states that we provide as input to the model. As can be seen from table 1, by providing prior knowledge to the model we are able to maintain fairly high accuracy for data with errors up to 50%.

Method	10	12	16
Baseline	0.065	0.078	0.086
Fuzzy data (25%)	0.090	0.153	0.094
Fuzzy data (50%)	0.135	0.160	0.131
Error (10%)	0.084	0.101	0.119
Error (20%)	0.114	0.136	0.148
Error (30%)	0.141	0.185	0.197
Error (40%)	0.182	0.230	0.237
Error (50%)	0.224	0.246	0.261
Error (60%)	0.236	0.274	0.283
Error (70%)	0.266	0.287	0.311
Error (80%)	0.283	0.292	0.330
Error (90%)	0.299	0.297	0.338

Table 2: The MAE of the prediction performed by our model in systems with delays, for 10, 12 and 16 variables.

5.4 Systems with Delays

To test the ability of our model to handle delays, we randomly generated 200 NLPs with delays up to $k = 5$, which are different from the training dataset, and performed experiments based on that. The results of the experiments are shown in table 2. All experiments include supplying the model with additional L_0 when the model fails to predict perfectly.

As evident from the results, our model is able to perform predictions under delayed setting. When we fuzz the data at about 25%, the prediction accuracy did not change by much, showing that the model is robust against fuzzy data. Also note that the model is still showing good prediction accuracy even when the error in the data is high. By providing additional background knowledge, the model is able to perform reasonable prediction even when the amount of error in the data is as high as 90%.

5.5 Discussion

Our model's main goal is to learn the logic program representation given a state transition. In figure 2, we show 2 separate logic program representations learned from 2 distinct state transitions that came from the fission benchmark. We can see that both of them share very similar features, and can confirm that our model did indeed manage to extract some high level features based on the input state transitions.

During the experiment above, we can confirm that providing background knowl-

(a) Logic program representation A

(b) Logic program representation B

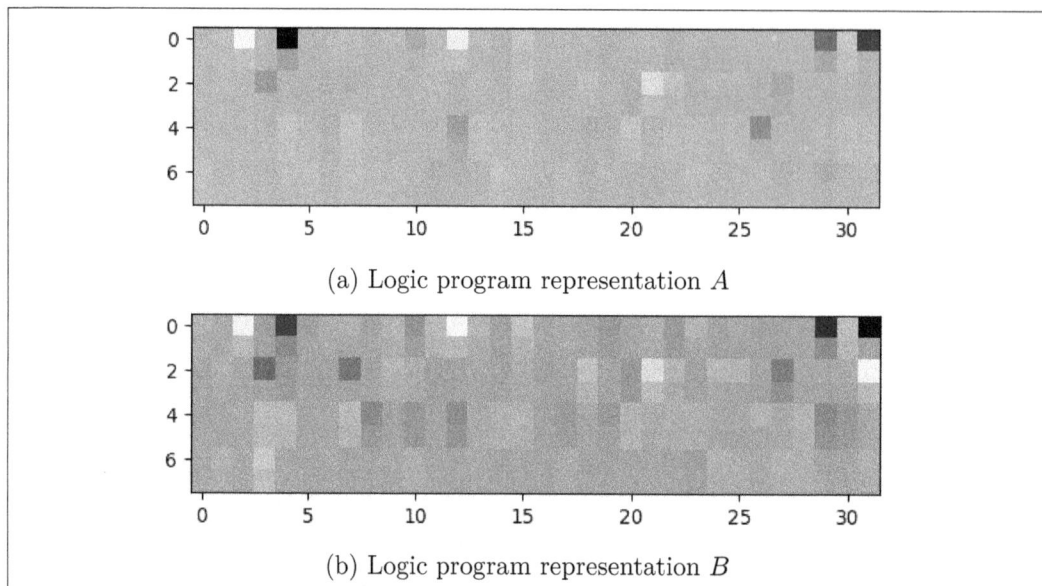

Figure 2: Two logic program representations learned from different state transitions from the fission benchmark

edge improves the model's prediction accuracy *most of the time*. There are times when adding background knowledge does not improve the prediction at all, such as when the transitions provided to obtain the background knowledge, already exists within the input transitions. This signifies a problem in which the background knowledge that we need to provide is not actually constructable by hand.

In addition, while performing the experiment, we trained separate models for dealing with NLPs with different number of variables. We did some preliminary testing and found that it is possible to train a model under larger number of variables, and then use it to predict systems with smaller number of variables. Since training separate models to work with different number of variables is extremely time consuming, this might open up the possibility of training the model just once and using it everywhere.

6 Conclusion

This paper's main contribution is the proposal of a method that allows us to provide background knowledge to the model in order to refine its predictions. This model works well in systems with delay setting, and can also deal well with fuzzy and erroneous data. In addition to that, this model also achieves an effect similar to one-shot learning, in which the model is given only 1 sample and it produces a good enough prediction result. Also, the representation learned can be transferred to a separate learning setting to further refine the results.

We performed experiments on 4 LFIT benchmarks, and showed that the model was able to perform the tasks at a similar level as its symbolic counterpart. We also showed that by providing background knowledge, the model was able to further refine its predictions and improve. We also tested the model on fuzzy and erroneous data, showing that the model is robust against both situation.

However there are still several shortcomings that we will like to address in future works. First, the output of the model being a matrix representation is not interpretable nor human-readable. We would like to be able to obtain the NLP in the symbolic form in order to interpret what the model was able to learn. Second, background knowledge supplied is limited in the form of either directly providing the matrix representation, or providing a separate state transition in order to obtain the matrix representation. Both of these methods are undesirable and we would like to be able to supply background knowledge in symbolic form.

References

[1] Abadi, M., Agarwal, A., Barham, P., et al.: TensorFlow: Large-scale machine learning on heterogeneous systems (2015), https://www.tensorflow.org/, software available from tensorflow.org

[2] Donoho, D.L.: High-dimensional data analysis: The curses and blessings of dimensionality. In: AMS Math Challenges Lecture. p. 132 (2000)

[3] d'Avila Garcez, A.S., Broda, K., Gabbay, D.M.: Symbolic knowledge extraction from trained neural networks: A sound approach. Artificial Intelligence 125(1), 155–207 (2001)

[4] d'Avila Garcez, A.S., Zaverucha, G.: The connectionist inductive learning and logic programming system. Applied Intelligence 11(1), 59–77 (1999)

[5] Inoue, K., Ribeiro, T., Sakama, C.: Learning from interpretation transition. Machine Learning 94(1), 51–79 (2014)

[6] Jennings, N.R., Sycara, K., Wooldridge, M.: A roadmap of agent research and development. Autonomous Agents and Multi-Agent Systems 1(1), 7–38 (Jan 1998), https://doi.org/10.1023/A:1010090405266

[7] Khan, A., Mandal, S., Pal, R.K., Saha, G.: Construction of gene regulatory networks using recurrent neural networks and swarm intelligence. Scientifica 2016 (2016)

[8] Martínez, D., Alenyà, G., Ribeiro, T., Inoue, K., Torras, C.: Relational reinforcement learning for planning with exogenous effects. Journal of Machine Learning Research 18(78), 1–44 (2017), http://jmlr.org/papers/v18/16-326.html

[9] Phua, Y.J., Tourret, S., Ribeiro, T., Inoue, K.: Learning logic program representation for delayed systems with limited training data. In: The 27th International Conference on Inductive Logic Programming (ILP 2017; OrlÃľans, France, September 4-6, 2017) (2017)

[10] Ribeiro, T., Inoue, K.: Learning prime implicant conditions from interpretation transition. In: Davis, J., Ramon, J. (eds.) Inductive Logic Programming. pp. 108–125. Springer International Publishing, Cham (2015)

[11] Ribeiro, T., Magnin, M., Inoue, K., Sakama, C.: Learning delayed influences of biological systems. Frontiers in Bioengineering and Biotechnology 2, 81 (2015)

[12] Ribeiro, T., Magnin, M., Inoue, K., Sakama, C.: Learning multi-valued biological models with delayed influence from time-series observations. In: 14th IEEE International Conference on Machine Learning and Applications, ICMLA 2015, Miami, FL, USA, December 9-11, 2015. pp. 25–31 (2015)

[13] Ribeiro, T., Tourret, S., Folschette, M., Magnin, M., Borzacchiello, D., Chinesta, F., Roux, O., Inoue, K.: Inductive learning from state transitions over continuous domains. In: Lachiche, N., Vrain, C. (eds.) Inductive Logic Programming. pp. 124–139. Springer International Publishing, Cham (2018)

[14] Tourret, S., Gentet, E., Inoue, K.: Learning human-understandable description of dynamical systems from feed-forward neural networks. In: Cong, F., Leung, A., Wei, Q. (eds.) Advances in Neural Networks - ISNN 2017. pp. 483–492. Springer International Publishing, Cham (2017)

Received 18 June 2018

COMPOSITIONALITY FOR RECURSIVE NEURAL NETWORKS

MARTHA LEWIS*
ILLC, University of Amsterdam
m.a.f.lewis@uva.nl

Abstract

Modelling compositionality has been a longstanding area of research in the field of vector space semantics. The categorical approach to compositionality maps grammar onto vector spaces in a principled way, but comes under fire for requiring the formation of very high-dimensional matrices and tensors, and therefore being computationally infeasible. In this paper I show how a linear simplification of recursive neural tensor network models can be mapped directly onto the categorical approach, giving a way of computing the required matrices and tensors. This mapping suggests a number of lines of research for both categorical compositional vector space models of meaning and for recursive neural network models of compositionality.

1 Introduction

Vector space semantics represents the meanings of words as vectors, learnt from text corpora. In order to compute the meanings of multi-word phrases and sentences, the principle of compositionality is invoked. This is that for a sentence $s = w_1 w_2 ... w_n$ there should be a function f_s that when applied to representations of the words w_i, will return a representation of the sentence s:

$$s = f_s(w_1, w_2, ... w_n)$$

One way to model meanings in a vector space is to use co-occurrence statistics [4]. The meaning of a word is identified with the frequency with which it appears near other words. A drawback of this approach is that antonyms appear in similar

---

Thanks to the organisers and participants of the NeSy2018 conference, and to the anonymous reviewers, for helpful comments and discussion.
*Funded by NWO Veni grant 'Metaphorical Meanings for Artificial Agents'

contexts and hence are indistinguishable. Another related difficulty is that vector spaces are notoriously bad for representing basic propositional logic. Nonetheless, the vector space model is highly successful in NLP. To model how words compose, a number of proposals have been made. These range from the simpler additive or multiplicative models given in [13] to full-blown tensor contraction models [7, 11]. In between is the Practical Lexical Function model of [14] which uses matrices to form function words such as adjectives and verbs.

The categorical compositional distributional model of [7] implements compositionality by mapping each grammatical type to a corresponding vector space. Grammatical reductions between types are modelled as linear maps between these vector spaces. Well-typed sentences reduce to vectors in the sentence space S. Vectors for nouns are learnt using cooccurrence statistics in corpora. Adjectives and verbs can be learnt using multilinear regression [1, 9], via a form of extensional composition [8], or by using techniques that reduce the size of the vector space [10].

Another way of building word meanings is via neural embeddings [12]. This strategy trains a network to predict nearby words by maximizing the probability of observing words in the neighbourhood of another. This is similar to the distributional idea, but rather than counting words, they are predicted. The prediction can happen in two directions: either a word is predicted from its context, called the continuous bag-of-words model, or the context is predicted from the word, called the skip-gram model. This method can then be extended to give a notion of compositionality. Recursive neural networks as used in [17] and [3] use a 'compositionality function' that computes the combination of two word vectors. This pairwise combination is applied recursively in a way that follows the parse structure of the phrase. The compositionality function has the structure of a feedforward neural network layer, possibly with additions such as a tensor layer. The parameters for the compositionality function and for the vectors themselves are trained using backpropagation.

The categorical approach maps nicely to formal semantics approaches. The role of verbs and adjectives as functions from the noun space to other spaces is clearly delineated. Words such as relative pronouns, whose meanings are not well modelled by distributional approaches, can be given a purely mathematical semantics. However, the representations of functional words soon become extremely large, so that learning, storing, and computing with these representations becomes infeasible. Another difficulty with this framework is that word types are fixed, so that there is no easy way to move between, say, noun meanings and verb meanings.

Neural network approaches in general do not have an explicit connection with formal semantics. In the case of recursive neural networks there is some connection, since the structure of the network respects the parse structure, but there is limited consideration of different grammatical types and how these might be used. Different grammatical types are all represented within the same vector space. Words that arguably have more of an 'information routing' function (such as pronouns, coordinators and so on) are also represented as vectors. However, these approaches are extremely successful. The word representations and the compositionality functions are more tractable than those of the full-blown tensor approach, and it is easy to consider a word vector as representing a number of different grammatical types - the same vector can be used to represent the noun 'bank' in 'financial bank' and the verb 'bank' in 'bank winnings'.

This paper shows how to understand a simplification of recursive neural networks within the categorical framework, namely, when the compositionality function is linear. Understanding recursive neural networks within this framework opens the door for us to use methods from formal semantics together with the neural network approach. I give an example of how we can express relative pronouns (words such as 'who') and reflexive pronouns ('himself') within the framework. This mapping also benefits the categorical approach. The high-order tensors needed for the categorical approach can be dispensed with, and word types can be made more fluid.

In the following, I firstly (section 2) give a description of categorical compositional vector space semantics. I go on to describe recursive neural networks and recursive neural tensor networks (section 3). In section 4 I show how linear recursive (tensor) networks can be given exactly the same structure as the categorical compositional model. Sections 5 and 6 outline the benefits of this analysis for each approach, and discuss how we can take the analysis further. In particular the possibility of reintroducing the non-linearity in recursive neural networks is considered.

2   Categorical Compositional Vector Semantics

In this section I describe elements of the category-theoretic compositional approach to meaning, as given in [7], and show the general method by which the grammar category induces a notion of concept composition in the semantic category. An introduction to the kind of category theory used here is given in [6]. The outline of the general programme is as follows [2]:

1. (a) Choose a compositional structure, such as a categorial grammar.

   (b) Interpret this structure as a category, the **grammar category**.

2. (a) Choose or craft appropriate meaning or concept spaces, such as vector spaces.

   (b) Organize these spaces into a category, the **semantics category**, with the same abstract structure as the grammar category.

3. Interpret the compositional structure of the grammar category in the semantics category via a functor preserving the type reduction structure.

4. This functor maps type reductions in the grammar category onto algorithms for composing meanings in the semantics category.

This paper describes one instantiation of this approach, using pregroup grammar and the category **FVect** of vector spaces and linear maps. This paper will use pregroup grammar, but it is also possible to use other approaches such as other categorial grammars, described in [5].

2.1 Pregroup grammar

The description of pregroup grammars given follows that of [15]. Whilst the details are slightly technical, the form of the grammar is very intuitive. Essentially we require a category that has types for nouns and for sentences, together with adjoint types, which are similar to inverses, a method for concatenating them, and morphisms that correspond to type reductions. The structure we desire for this category is termed *compact closed*. Details are given in [7] and [15].

The category G used for grammar is roughly as follows. The grammar is built over a set of types. We consider the set containing just n for noun and s for sentence. Each type has two adjoints x^r and x^l. Complex types can be built up by concatenation of types, for example $x \cdot y^l \cdot z^r$, and we often leave out the dot so $xy = x \cdot y$. There is also a unit type such that $x1 = 1x = x$. Types and their adjoints interact via the following morphisms:

$$\epsilon_x^r : x \cdot x^r \to 1, \qquad \epsilon_x^l : x^l \cdot x \to 1$$
$$\eta_x^r : 1 \to x^r \cdot x, \qquad \eta_x^l : 1 \to x \cdot x^l$$

The morphisms ϵ_x^r and ϵ_x^l can be thought of as *type reduction* and the morphisms η_x^r and η_x^l can be thought of as *type introduction*. A string of grammatical types $t_1, ...t_n$ is then said to be grammatical if it reduces, via the morphisms above, to the sentence type s.

Example 1. Consider the sentence *'dragons breathe fire'*. The nouns *dragons* and *fire* are of type n, and the verb *breathe* is given the type $n^r s n^l$. *'dragons breathe fire'* therefore has type $n(n^r s n^l)n$. Then we have the following type reductions:

$$(\epsilon_n^r \cdot id_s \cdot \epsilon_n^l)(n(n^r s n^l)n) = (\epsilon_n^r \cdot id_s \cdot \epsilon_n^l)((nn^r)s(n^l n))$$
$$\to (id_s \cdot \epsilon_n^l)s(n^l n) \to s$$

The above reduction can be given a neat graphical interpretation as follows:

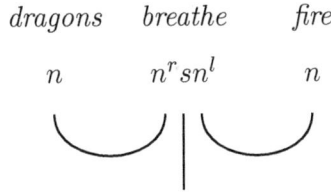

This diagrammatic calculus is fully explained in [7], amongst others. Essentially we can think of the u-shaped 'cups' as type reductions, and calculations can be made by manipulating the diagrams as if they lie on a flat plane, maintaining numbers of inputs and outputs.

2.2 Mapping to vector spaces

We use the category **FVect** of finite dimensional vector spaces and linear maps, which is also compact closed. We describe a functor $\mathcal{F} : G \to$ **FVect** that maps the noun type n to a vector space N, the sentence type s to S, the unit 1 to $\mathbb{R}$, concatenation maps to $\otimes$, i.e., the tensor product of vector spaces, adjoints are lost, ϵ_p^r and ϵ_p^l map to tensor contraction, and η_p^r and η_p^l map to identity maps.

Example 2. Consider again the sentence *'dragons breathe fire'*. The nouns *dragons* and *fire* have type n and so are represented in some vector space N of nouns. The transitive verb *breathe* has type $n^r s n^l$ and, hence, is represented by a vector in the vector space $N \otimes S \otimes N$ where S is a vector space modelling sentence meaning. The meaning of *'dragons breathe fire'* is the outcome of applying the type reduction morphisms given in

$$\epsilon_N \otimes 1_S \otimes \epsilon_N : N \otimes (N \otimes S \otimes N) \otimes N \to S \tag{1}$$

i.e. sequences of tensor contractions, to the product

$$\overrightarrow{dragons} \otimes \overrightarrow{breathe} \otimes \overrightarrow{fire} \tag{2}$$

This nicely illustrates the general method. The meaning category supplies vectors for *dragons*, *breathe*, and *fire*. The grammar category then tells us how to stitch these together. The essence of the method should be thought of as the diagram

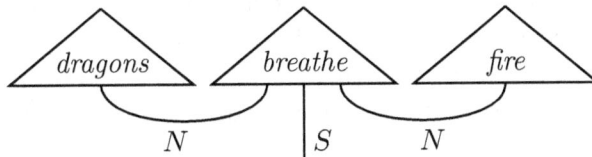

713

where we think of the words as meaning vectors (2) and the wires as the map (1). Again, the 'cups' can be thought of as type reductions. Linear-algebraically, the map (1) and the diagram above are equivalent to the following. Suppose we have a set of basis vectors $\{\vec{e}_i\}_i$. Define

$$\overrightarrow{dragons} = \sum_i d_i \vec{e}_i, \qquad \overrightarrow{breathe} = \sum_{ijk} b_{ijk} \vec{e}_i \otimes \vec{e}_j \otimes \vec{e}_k, \qquad \overrightarrow{fire} = \sum_i f_i \vec{e}_i$$

Then

$$\overrightarrow{dragons\ breathe\ fire} = (\epsilon_N \otimes 1_S \otimes \epsilon_N)\overrightarrow{dragons} \otimes \overrightarrow{breathe} \otimes \overrightarrow{fire}$$

$$= (\epsilon_N \otimes 1_S \otimes \epsilon_N)\left(\sum_i d_i \vec{e}_i \otimes \sum_{jkl} b_{jkl} \vec{e}_j \otimes \vec{e}_k \otimes \vec{e}_l \otimes \sum_m f_m \vec{e}_m\right)$$

$$= (1_S \otimes \epsilon_N)\left(\sum_{ijkl} d_i b_{jkl} \vec{e}_j \otimes \vec{e}_k \otimes \sum_m f_m \vec{e}_m\right) = \sum_{ijk} d_i b_{ijk} f_k \vec{e}_j$$

where this last expression is a single vector in the sentence space.

3   Neural Network Models

Neural networks are used both as a way of building meaning vectors and as a way of modelling compositionality in meaning spaces. Mikolov *et al.* [12] describes a pair of methods that build vectors by using context windows, and making predictions about the likely content of either the context window or the word itself. Phrases and sentences are represented in the same space as the words. To compute vectors for multi-word sentences and phrases, [17] use tree-structured recursive neural networks. The phrases and sentences output by the network can then be used for various tasks, notably sentiment analysis. The sections below summarise recursive neural networks and recursive tensor neural networks. In the following sections we assume that words are represented as vectors in $\mathbb{R}^n$.

3.1   Recursive neural networks

Recursive neural networks (TreeRNNs) have a tree-like structure. When applied to sentences, the tree represents the syntactic structure of the sentence. A schematic of a recursive neural network is given in Figure 1. The words of a sentence are represented as vectors. Words can be combined via the *compositionality function* g to form a parent vector. In the networks we discuss here, the parent vectors are

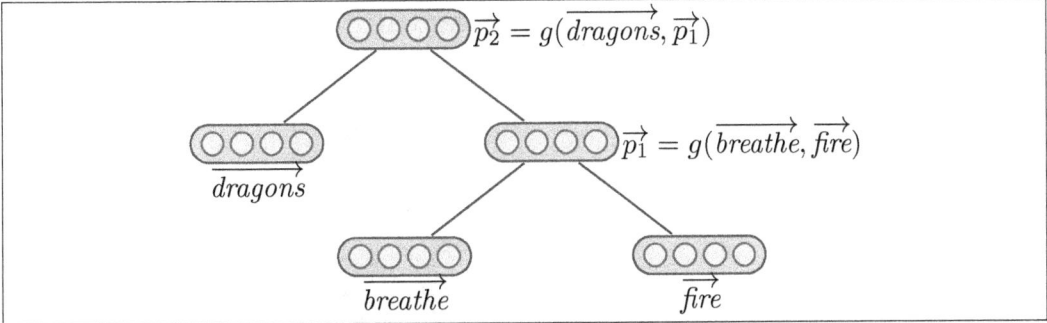

Figure 1: Schematic of an TreeRNN. Word vectors and/or parent vectors p_i are combined using the compositionality function g according to the parse tree. The vector $\vec{p_1}$ corresponds to the verb phrase *breathe fire* and the vector $\vec{p_2}$ corresponds to the whole sentence *dragons breathe fire*.

of the same dimensionality as the input vectors, meaning that the compositionality function can be applied recursively according to the parse tree. The compositionality function and the input vectors themselves are learnt by error backpropagation.

The compositionality function for a TreeRNN is usually of the form

$$g_{TreeRNN} : \mathbb{R}^n \times \mathbb{R}^n \to \mathbb{R}^n :: (\vec{v_1}, \vec{v_2}) \mapsto f_1 \left(M \cdot \begin{bmatrix} \vec{v_1} \\ \vec{v_2} \end{bmatrix} \right)$$

where $\vec{v_i} \in \mathbb{R}^n$, $\begin{bmatrix} - \\ - \end{bmatrix}$ is vertical concatenation of column vectors, $M \in \mathbb{R}^n \otimes \mathbb{R}^{2n}$, and $(- \cdot -)$ is tensor contraction. f_1 is a squashing function that is applied pointwise to its vector argument, for example $f = \tanh$. The parent vector that forms the root of the tree is the representation of the whole sentence. Parent vectors within the tree represent subphrases of the sentence. The matrix M and the input vectors are learnt during training.

3.2 Recursive neural tensor networks

Recursive neural tensor networks (TreeRNTNs) are similar to TreeRNNs but differ in the compositionality function g. The function g is as follows:

$$g_{TreeRNTN} : \mathbb{R}^n \times \mathbb{R}^n \to \mathbb{R}^n :: (\vec{v_1}, \vec{v_2}) \mapsto g_{TreeRNN}(\vec{v_1}, \vec{v_2}) + f_2 \left(\vec{v_1}^\top \cdot T \cdot \vec{v_2} \right)$$

where $\vec{v_i}$ and $(- \cdot -)$ are as before, $T \in \mathbb{R}^n \otimes \mathbb{R}^n \otimes \mathbb{R}^n$ and f_2 is a squashing function. Again, the input vectors, matrix M and tensor T are learnt during training.

4 Mapping between categorical and TreeRNN compositionality

It is now possible to model a simplifed version of TreeRNNs within the categorical vector space semantics of [7]. I show show how a linearized version can be modelled within **FVect** using pregroup grammar as the grammar category.

With a (drastic) simplification of the compositionality function $g_{TreeRNTN}$ there is an immediate correspondence between the TreeRNTN model and a simplified version of the categorical model. We drop both the non-linearity and the matrix part of the function g, giving:

$$g_{Lin} : \mathbb{R}^n \times \mathbb{R}^n \to \mathbb{R}^n :: (\vec{v_1}, \vec{v_2}) \mapsto \left(\vec{v_1}^\top \cdot T \cdot \vec{v_2}\right)$$

Now the tensor T is just a multilinear map, i.e., morphism in **FVect**, and we can therefore describe a direct translation between linear TreeRNTNs and categorical compositional vector space semantics with pregroups.

Recall that in the categorical model we had a nice diagrammatic calculus to represent the calculations we were making. We also had a schematic for the TreeRNNs. With the simplified compositionality function, we can translate that schematic into the diagrammatic calculus, shown in figures 2, 3, and 4.

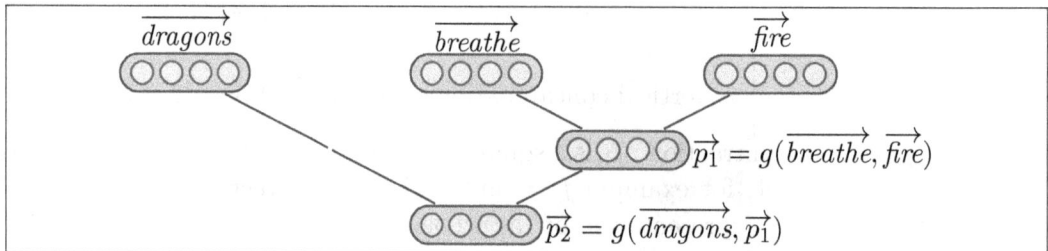

Figure 2: The TreeRNN schematic turned upside down and one edge lengthened

These diagrams show how the interior of the verb has been has been analysed into two instances of the compositionality function wired together, with the verb vector $\overrightarrow{breathe}$ as input. This means that rather than learn large numbers of parameters for each word in the lexicon, just one tensor comprising the compositionality function needs to be learnt, together with vectors in N for each word. This mapping can be carried out for other parts of speech. The representations of adjectives and intransitive verbs are given in figures 6 and 7, each requiring just one instance of the compositionality function. In section 5.2, we discuss how we can analyze other sorts of words such as relative pronouns and reflexive pronouns.

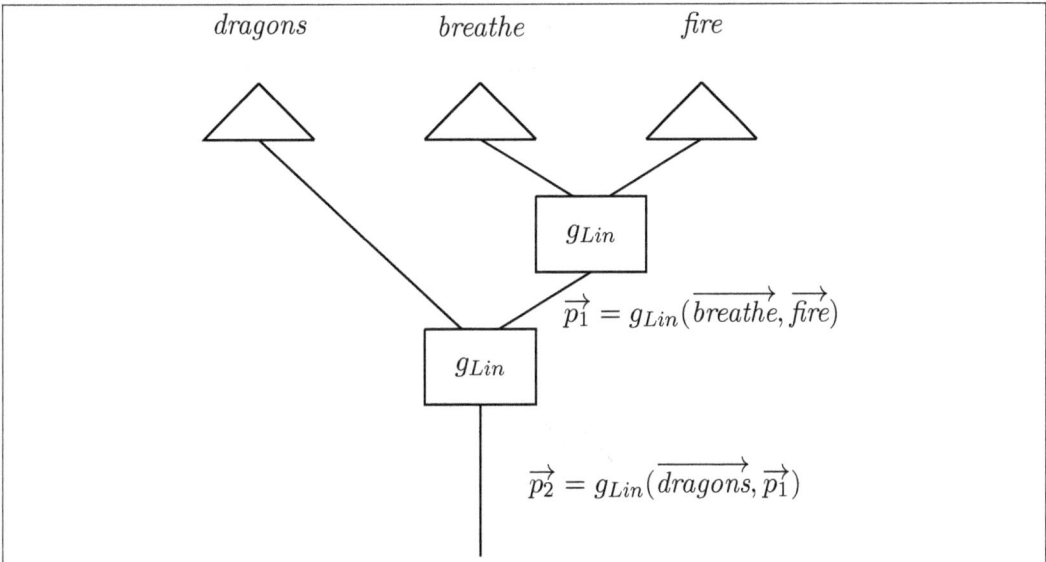

Figure 3: The schematic translated into the diagrammatic calculus. The compositionality function g_{Lin} is just a tensor with no nonlinearity applied.

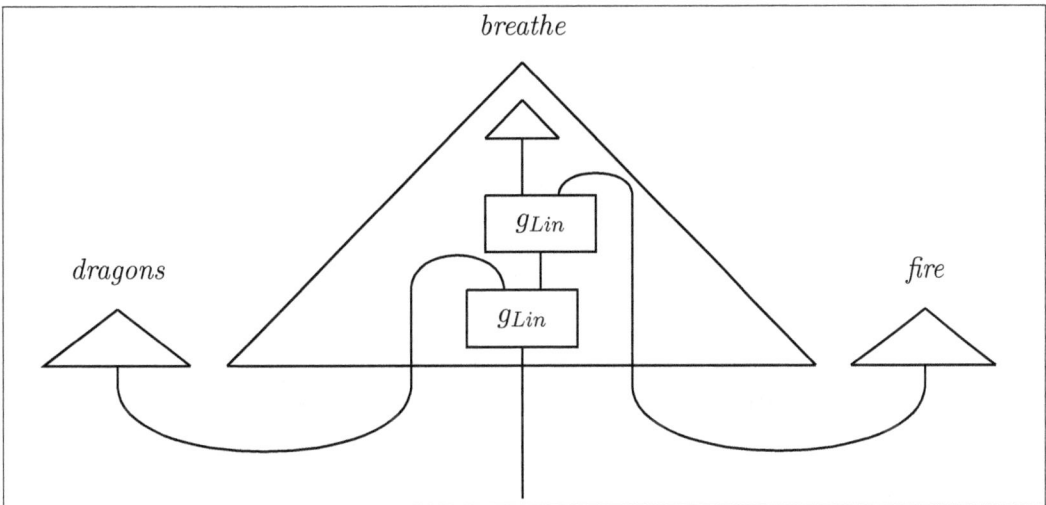

Figure 4: The diagram in figure 3 with wires bent. This is allowed since we are now working in the category **FVect**.

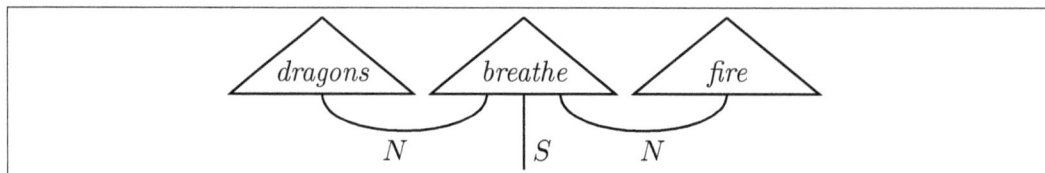

Figure 5: We can therefore see the case in 4 as an instance of the categorical method, where the interior of the tensor is created using two instances of the compositionality function g_{Lin}

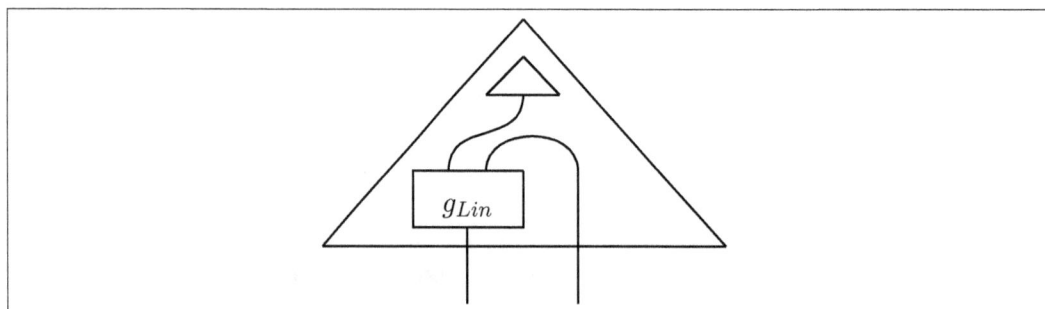

Figure 6: Adjective formed from part of a TreeRNTN.

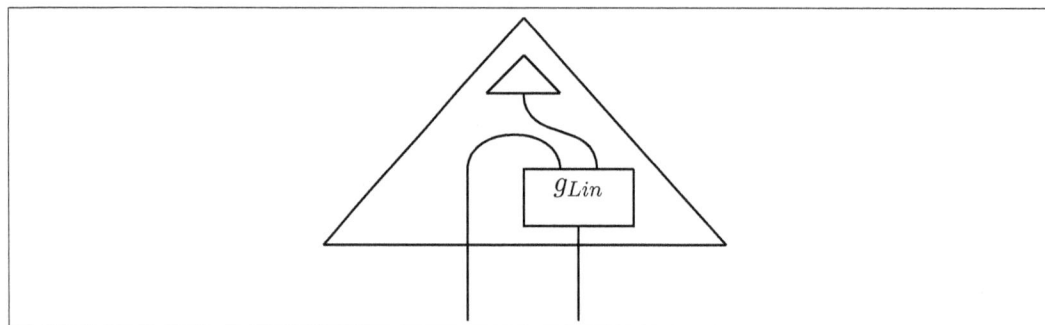

Figure 7: Intransitive verb formed from part of an TreeRNTN.

5   Benefits

In outlining this comparison a number of benefits arise. This section outlines benefits for the categorical model and then for RNN models.

5.1 Categorical models

One of the main charges levelled at the categorical compositional distributional semantics mode is that the dimensionality of the tensors required is too high, and that training is too expensive. The correspondence I have outlined here gives an approach where the number of high-dimensional tensors to train is limited.

In the simplest case, one linear compositionality function could be learned, together with vectors for each word. The learning algorithm for this approach will be similar to strategies used for training recursive neural networks. The networks will therefore be as easy, or easier, to train than the TreeRNNs used by [17] and [3]. However, since the compositionality functions to train are now linear, the results obtained are unlikely to be as good as those obtained using full TreeRNNs. One strategy to alleviate this is as follows. Different compositionality functions could be used for different word types. So, for example, we would have functions g_{adj} for an adjective, g_{iv} and g_{tv} for a transitive verb. For example, the functions for the adjective is $g_{adj}(\overrightarrow{v}_n) = \overrightarrow{v}_a^\top T_{adj} \overrightarrow{v}_n$, and for an intransitive verb is $g_{iv}(\overrightarrow{v}_n) = \overrightarrow{v}_n^\top T_{iv} \overrightarrow{v}_i$, where $\overrightarrow{v}_a$ is the vector of the adjective, $\overrightarrow{v}_i$ is the vector of the intransitive verb, and $\overrightarrow{v}_n$ is the vector of the noun. Using this strategy, the noun space and the sentence space can now be separated so that sentences no longer have to inhabit the same space as nouns.

A further benefit for the categorical model is that this approach alleviates the brittleness of the representations learnt. Rather than learning individual tensors for each functional word, we are simply learning a small number of compositionality functions. This means that we can switch between the noun 'bank' and the verb 'bank' simply by plugging the word vector *bank* into the relevant function.

Furthermore, since this approach is a simplification of the model of [7], extensions of that model can also be applied. In particular, information-routing words like relative pronouns can be modelled using the approaches outlined in [16]. This is discussed further in the next section.

5.2 TreeRNN models

Although TreeRNNs have fewer parameters and more flexibility than the categorical vector space models, the compositional mechanism they use is 'one size fits all'. The TreeRNN approach as elaborated so far does not distinguish between content words such as 'dog', 'brown', and information routing words such as pronouns and logical words. The approach outlined here makes an explicit connection between formal semantics approaches in the form of pregroup grammars on the one hand, and neural network approaches for composition on the other. This means that we can use strategies from formal semantics to represent the meaning of information routing words. The benefit of doing so is two-fold. Firstly, it may improve training time, since the compositionality function will not have to encompass this aspect of composition. Secondly, by separating out some of the compositional mechanism, we make the behaviour of the network more transparent. The roles of certain words will be modelled as functions that do not need to be learnt. I give below two examples: relative pronouns as analysed in [16] and reflexive pronouns.

5.2.1 Relative pronouns

[16] analyze relative pronouns by using the Frobenius algebra structure available on finite-dimensional vector spaces. Full details of how Frobenius algebras are defined and used are given in those papers, but briefly, we can consider these to introduce copying, merging, and deleting mechanisms into the semantics.

In **FVect**, any vector space V with a fixed basis $\{\overrightarrow{e_i}\}_i$ has a Frobenius algebra over it, explicitly given by:

$$\Delta :: \overrightarrow{e_i} \mapsto \overrightarrow{e_i} \otimes \overrightarrow{e_i} \qquad\qquad \iota :: \overrightarrow{e_i} \mapsto 1 \qquad (3)$$
$$\mu :: \overrightarrow{e_i} \otimes \overrightarrow{e_i} \mapsto \overrightarrow{e_i} \qquad\qquad \zeta :: 1 \mapsto \overrightarrow{e_i} \qquad (4)$$

Linear-algebraically, the Δ morphism takes a vector and embeds it into the diagonal of a matrix. The μ morphism takes a matrix $z \in W \otimes W$ and returns a vector consisting only of the diagonal elements of z. If the matrix z is the tensor product of two vectors $z = \overrightarrow{v} \otimes \overrightarrow{w}$, then $\mu(v \otimes w) = v \odot w$ where $(- \odot -)$ corresponds to pointwise multiplication. These operations extend to higher-order tensors.

In pregroup grammar, the word 'who' is given the type $n^r n s^l n$. Rather than learn parameters for an order 4 tensor, [16] show how it can be given a purely mathematical meaning. This is shown diagrammatically below:

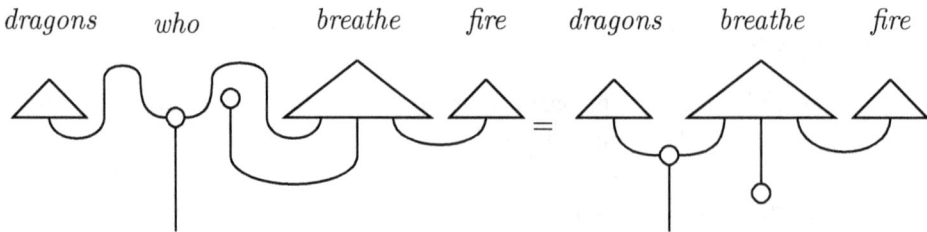

The word 'who' is equivalent to discarding the sentence part of the verb and pointwise multiplying the vectors for *dragons* and *breathe fire*.

5.2.2 Reflexive Pronouns

Reflexive pronouns are words such as 'himself'. These words also have an information routing role. In a sentence like *John loves himself*, we want the content of *John* to be copied out and routed to the object of the verb. The pregroup type of the pronoun 'himself' can be given as $ns^r n^{rr} n^r s$. We can give the reflexive pronoun a purely mathematical semantics as follows:

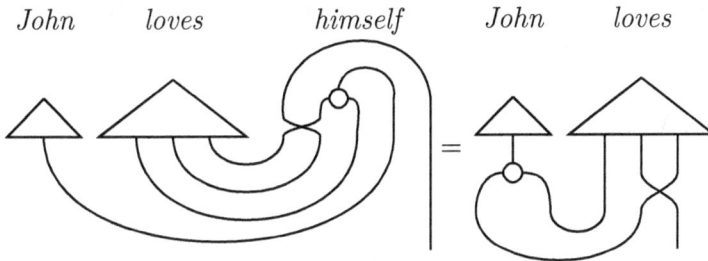

The reflexive pronoun takes in the noun, copies it, and plugs it into both the subject and the object of the verb, and returns the resulting sentence.

This treatment of reflexive and relative pronouns is part of a larger programme, relating vector space models of meaning and formal semantics. The idea is that some words can be thought of as 'information routing' - they move information around a sentence, and at least part, if not all, of their meaning should be purely mathematical. In contrast, information-carrying words like nouns and adjectives, have meaning determined by co-occurrence, rather than by a mathematical function. In the TreeRNN approach, this distinction is not made, meaning that the compositionality function learnt must take into account both statistical and information-routing kinds of meaning. The proposal here is that information-routing words can be understood as part of the structure of the tree, rather than as vectors.

6 Conclusions and Further Work

The aim of this paper is to set out a mapping between the categorical compositional vector space semantics of [7] and the recursive neural network (TreeRNN) models of [17] and [3]. I have shown how a linear version of TreeRNNs can be modelled directly within the categorical model. This gives a strategy for simplifying the training for the categorical model, and also means that the categorical model is more flexible in its word representations. Since a linearized neural network is not going to be as successful as a standard network, I have also suggested learning individual networks for individual grammatical types, as a way of improving performance whilst still requiring many fewer parameters than the standard categorical model. Modelling TreeRNNs within the categorical framework means that we can use ideas from formal semantics to simplify networks. I have shown how relative pronouns and reflexive pronouns can be analysed as having a purely mathematical semantics. This means that the networks learnt do not need to take this sort of compositionality into account. Furthermore, using the purely mathematical semantics when available means that the networks are more transparent. With analysis of these words, the compositionality function learnt can specialise to contentful words, rather than information routing words.

6.1   Further work

Section 4 showed how we can express a linear version of TreeRNTNs within the categorical compositional vector space model. However, using only linear transformations limits what these networks can do. Ongoing work is to examine how non-linearity can be reintroduced, by changing the categorical framework in which we work. The most promising avenue seems to be to change to monoidal biclosed categories and Lambek categorial grammar.

There are a number of other avenues for further work to be considered. On the implementation side:

- The performance of linear TreeRNNs can be tested against the usual categorical apporaches to learning words.

- The performance of linear TreeRNNs with specialised word-type networks can be tested against standard TreeRNNs.

- The performance of TreeRNNs with formally analyzed information-routing words can be tested.

- The effects of switching between word types can be investigated.

On the theoretical side:

- The analysis of reflexive pronouns can be extended to other pronouns and anaphora.

- Investigating meanings of logical words and quantifiers.

- Extending the analysis to other types of recurrent neural network such as long short-term memory networks or gated recurrent units.

References

[1] Marco Baroni and Roberto Zamparelli. Nouns are vectors, adjectives are matrices: Representing adjective-noun constructions in semantic space. In *Proceedings of the 2010 Conference on Empirical Methods in Natural Language Processing*, pages 1183–1193. Association for Computational Linguistics, 2010.

[2] Joe Bolt, Bob Coecke, Fabrizio Genovese, Martha Lewis, Dan Marsden, and Robin Piedeleu. Interacting conceptual spaces i: Grammatical composition of concepts. *arXiv preprint arXiv:1703.08314*, 2017.

[3] Samuel R Bowman and Christopher Potts. Recursive neural networks can learn logical semantics. *ACL-IJCNLP 2015*, page 12, 2015.

[4] J.A. Bullinaria and J.P. Levy. Extracting semantic representations from word co-occurrence statistics: A computational study. *Behavior research methods*, 39(3):510–526, 2007.

[5] B. Coecke. An alternative gospel of structure: order, composition, processes. In C. Heunen, M. Sadrzadeh, and E. Grefenstette, editors, *Quantum Physics and Linguistics. A Compositional, Diagrammatic Discourse*, pages 1–22. Oxford University Press, 2013.

[6] B. Coecke and E.O. Paquette. Categories for the practising physicist. In *New Structures for Physics*, pages 173–286. Springer, 2011. doi: 10.1007/978-3-642-12821-9.

[7] B. Coecke, M. Sadrzadeh, and S. Clark. Mathematical foundations for a compositional distributional model of meaning. *Linguistic Analysis*, 36:345–384, 2010.

[8] Edward Grefenstette and Mehrnoosh Sadrzadeh. Experimental support for a categorical compositional distributional model of meaning. In *Proceedings of the Conference on Empirical Methods in Natural Language Processing*, pages 1394–1404. Association for Computational Linguistics, 2011.

[9] Edward Grefenstette, Georgiana Dinu, Yao-Zhong Zhang, Mehrnoosh Sadrzadeh, and Marco Baroni. Multi-step regression learning for compositional distributional semantics. *arXiv preprint arXiv:1301.6939*, 2013.

[10] Dimitri Kartsaklis, Mehrnoosh Sadrzadeh, and Stephen Pulman. A unified sentence space for categorical distributional-compositional semantics: Theory and experiments. In *Proceedings of COLING 2012: Posters*, pages 549–558, 2012.

[11] Jean Maillard, Stephen Clark, and Edward Grefenstette. A type-driven tensor-based semantics for ccg. *EACL 2014*, page 46, 2014.

[12] Tomas Mikolov, Ilya Sutskever, Kai Chen, Greg S Corrado, and Jeff Dean. Distributed representations of words and phrases and their compositionality. In *Advances in neural information processing systems*, pages 3111–3119, 2013.

[13] J. Mitchell and M. Lapata. Composition in distributional models of semantics. *Cognitive science*, 34(8):1388–1429, 2010.

[14] Denis Paperno, Marco Baroni, et al. A practical and linguistically-motivated approach to compositional distributional semantics. In *Proceedings of the 52nd Annual Meeting of the Association for Computational Linguistics (Volume 1: Long Papers)*, volume 1, pages 90–99, 2014.

[15] A. Preller and M. Sadrzadeh. Bell states and negative sentences in the distributed model of meaning. *Electronic Notes in Theoretical Computer Science*, 270(2):141–153, 2011. doi: 10.1016/j.entcs.2011.01.028.

[16] M. Sadrzadeh, S. Clark, and B. Coecke. The Frobenius anatomy of word meanings I: subject and object relative pronouns. *Journal of Logic and Computation*, 23:1293–1317, 2013. arXiv:1404.5278.

[17] Richard Socher, Alex Perelygin, Jean Wu, Jason Chuang, Christopher D Manning, Andrew Ng, and Christopher Potts. Recursive deep models for semantic compositionality over a sentiment treebank. In *Proceedings of the 2013 conference on empirical methods in natural language processing*, pages 1631–1642, 2013.

Received 18 June 2018

TOWARDS FUZZY NEURAL CONCEPTORS

TILL MOSSAKOWSKI
Otto-von-Guericke Universität Magdeburg, Germany

MARTIN GLAUER
Otto-von-Guericke Universität Magdeburg, Germany

RĂZVAN DIACONESCU
Simion Stoilow Institute of Mathematics of the Romanian Academy, Bucharest, Romania

Abstract

Conceptors are an approach to neuro-symbolic integration based on recurrent neural networks. Jaeger has introduced two-valued logics for conceptors. We observe that conceptors are essentially fuzzy in nature, and hence develop a fuzzy subconceptor relation and a fuzzy logic for conceptors.

1 Introduction

Neural networks have been successfully used for learning tasks [14], but they exhibit the problem that the way they compute their output generally cannot be interpreted or explained at a higher conceptual level [15]. The field of neuro-symbolic integration [2] addresses this problem by combining neural networks with logical methods. However, most approaches in the field (like e.g. logic tensor networks [4]) are localist, that is, predicates or other symbolic items are represented in small sub-networks. This contrasts with the distributed representation of knowledge in (deep learning) neural networks, which seems to be much more flexible and powerful.

Jaeger's conceptors [10, 11, 12, 8] provide such a distributed representation while simultaneously providing logical operators and concept hierarchies that foster explainability. The basic idea is to take a recurrent neural network and not use it for learning through back-propagation, but rather as a reservoir, i.e. a network with fixed randomly created connection weights. The reservoir feed it with input signals,

We thank the anonymous referees for valuable comments.

Vol. 6 No. 4 2019

leading to a state space that can be captured as a certain ellipsoid using a conceptor matrix. Conceptor matrices are positive semi-definite matrices with singular values (which represent the lengths of the ellipsoid axes) ranging in [0,1]. Details will be given below.

In [11], Jaeger introduces and studies algebra of conceptors, providing the quasi-Boolean operations "and", "or" and "not" (which however satisfy only part of the laws of Boolean algebra), as well as a scaling operation called aperture adaption, and an interpolation operation. A crucial advantage of conceptors over ordinary neural networks is that using the algebra of conceptors, training examples can easily be added to conceptors, without the need of re-training with the whole sample. This also has been applied to deep learning [8] in order to avoid catastrophic forgetting. Moreover, the Löwner ordering on conceptor matrices expresses a concept hierarchy. For reasoning about conceptors, two logics are introduced, an extrinsic and an intrinsic one.

We here argue that both of these logics are not adequate for reasoning about conceptors, because they both can ultimately speak only about the Löwner ordering, i.e. crisp statements that can be either true or false. We propose that a more promising approach is to view conceptors as a kind of fuzzy sets. Indeed, their quasi-Boolean operators satisfy the (appropriate generalisation of) T-norm and T-conorm laws, and form a (generalised) De Morgan Triplet [18, 7]. This is remarkable, because conceptors have not been introduced as a neuro-fuzzy approach (and note that neuro-fuzzy approaches generally are localist in the above sense, while conceptors provide a global distributed representation of knowledge).

We argue that an appropriate conceptor logic should not have crisp but fuzzy statements as its atomic constituents. The main reason is that conceptors are built from signals, which typically are noisy to some degree. Hence, we expect that a subconceptor relation should be immune to small disturbances and noise in signals. This cannot be achieved with the Löwner ordering. Hence, we introduce a fuzzy variant of the Löwner ordering that can be used as a fuzzy subconceptor relation. This is also in accordance with the finding that conceptors behave like fuzzy sets.

Our proposed conceptor logic hence includes all what Jaeger has in his logic: formation of conceptors from signals, the quasi-Boolean logical operations, aperture scaling and interpolation. This is extended with a fuzzy subconceptor relation, as well as a classification of signals (which can be seen as fuzzy set membership). On this basis, we develop a fuzzy logic for conceptors. Concept hierarchies can be obtained using hierarchical clustering. Note that our approach is different from combinations of machine learning and fuzzy methods in the literature [9], where the basic relations that are learned are fuzzified. By contrast, our approach incorporates fuzzy logic *on top of* neural networks. This is also the reason why we use non-fuzzy

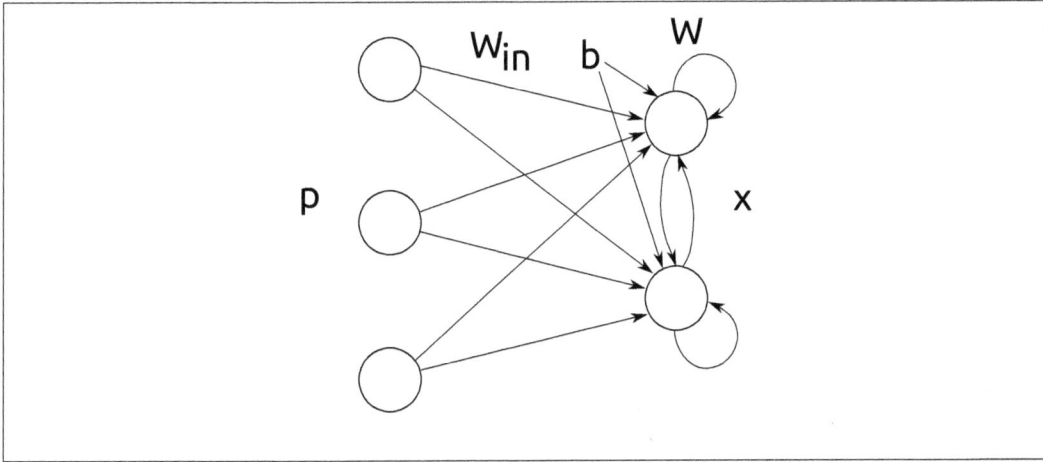

Figure 1: Structure of the recurrent neural network

hierarchical clustering.

2 Conceptors

In this section, we recall some basic notions and definitions from [11]. Conceptors arise from observing the dynamics of randomly created recurrent neural networks (RNNs). The latter are also called reservoirs. An input signal p drives this network. For time steps $n = 0, 1, 2, \ldots L$, the system is governed by the equation

$$x(n+1) = tanh(W\,x(n) + W^{in}p(n+1) + b)$$

see also Fig. 1. Here,

- $p(n)$ is a K-dimensional input vector at time step n

- $x(n)$ is the N-dimensional state vector at time step n

- W is an $N \times N$ matrix of reservoir-internal connection weights

- W^{in} is an $N \times K$ vector of input connection weights

- b is a $N \times 1$ bias vector

- $tanh$ is the hyperbolic tangent.

Note that W, W^{in} and b are randomly created.

If such a network of dimension N is observed for L steps, the state vectors $x(0), \ldots x(L)$ can be collected into an $N \times L$ matrix X representing the cloud of visited states in the N-dimensional reservoir state space. From this, the reservoir state correlation matrix is computed as

$$R = XX^T / L$$

This is comparable to a covariance matrix in statistics. R_{ij} is a measure for how much the $x(n)_i$ relate to the $x(n)_j$ over all time steps $n = 0, 1, 2, \ldots L$.

2.1 From Correlation Matrices to Conceptors

A conceptor C is a normalised ellipsoid (inside the unit sphere) that (like the state correlation matrix R) represents the cloud of visited states. It can be computed from state correlation matrix R via the following normalisation:

$$C(R, \alpha) = R(R + \alpha^{-2}I)^{-1} = (R + \alpha^{-2}I)^{-1}R \tag{1}$$

where $\alpha \in (0, \infty)$ is a scaling factor called *aperture*. Conversely, given a conceptor C, the state correlation matrix can be computed via

$$R(C, \alpha) = \alpha^{-2}(I - C)^{-1}C = \alpha^{-2}C(I - C)^{-1}$$

A conceptor forms an ellipsoid that can be seen as a "fingerprint" of the activity the RNN. It can also be seen as a polarisation filter that filters out certain dimensions (which means that $Cx \approx 0$ for vectors x that live only in these dimensions) and leaves through others (which means that $Cx \approx x$ for vectors x that live only in these dimensions). Fig. 2 shows four sample input signals, their behaviour on two neurons, as well as the singular values of the state correlation matrix R and those of the conceptor C. The singular value decomposition (SVD) of R and C can be seen as a principal component analysis, where the singular values represent the different relevant dimensions of the ellipsoid. While the singular values of R are arbitrary, those of C are not. Equation (1) always leads to matrices with normalised singular values. This leads to:

Definition 1 ([11]). *Given a dimension $M \in \mathbb{N}$, a conceptor is a positive semi-definite real-valued $M \times M$-matrix with singular values ranging in [0,1].*

Recall that an eigenvalue of a matrix M is a value $\lambda \in \mathbb{R}$ such that $Mv = \lambda v$ for some vector v, and that a matrix is positive semi-definite iff it is symmetric and all

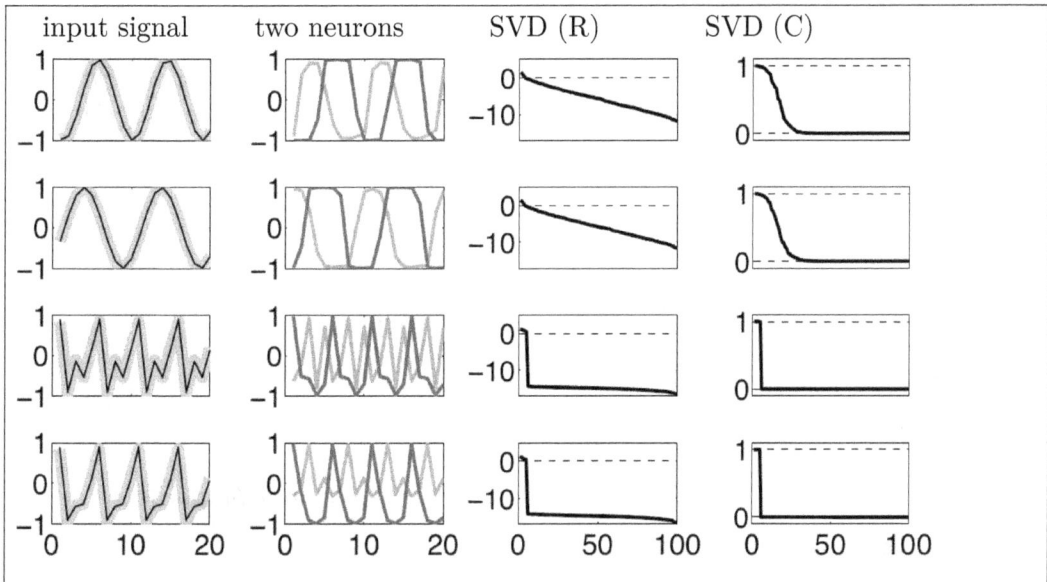

Figure 2: Different input signals (two sine and two sawtooth waves) and the singular value decompositions of their state correlation matrices R and their conceptors C. Image from [11]

of its eigenvalues are non-negative. Moreover, for such matrices, singular values and eigenvalues coincide. Hence, a conceptor can be equivalently defined as a symmetric matrix whose eigenvalues range in [0,1].

A conceptor is *hard*, if all singular values are in $\{0, 1\}$. A conceptor is a *conception vector*, if at most the diagonal is nonzero [11]. Jaeger shows that with an embedding into a higher dimension, conception vectors (in this higher dimension) can have capabilities resembling those of conceptors (in the lower dimension).

2.2 Quasi-Boolean Operations on Conceptors

A crucial feature of conceptors that logical and other operations are available for them. Logical disjunction can be used for adding more training examples gradually, without catastrophic forgetting (known from deep learning approaches) taking place, see [8] and Sect. 2.3 below (where also negation is used). Aperture adaption can change the scaling parameter α discussed above, and is needed to keep conceptors between de-differentiation and over-excitement (see [11] for explanations). Interpolation can be used to smoothly move from one conceptors to another one. Jaeger illustrates this with motions of a human skeleton, where walking is smoothly transformed into running, dancing or sitting etc[1].

Definition 2 (Quasi-Boolean operations [11]). *On conceptors, we can define the following quasi-Boolean operations, as well as operations of aperture adaption and interpolation:*

- $\neg C := 1 - C$

- $C \vee D := C(R(C,1) + R(D,1), 1)$

- $C \wedge D := \neg(\neg C \vee \neg D)$

- $\bot = 0$ *(zero matrix, plays the role of* false*)*

- $\top = I$ *(identity matrix, plays the role of* true*)*

- $\varphi(C, \gamma) = C/(C + \gamma^{-2}(I - C))$ *for $0 < \gamma < \infty$ (aperture adaption)*

- $\beta_b(C, B) = bC + (1 - b)B$ *(interpolation)*

In [11], it is shown that these quasi-Boolean operations satisfy some useful laws like associativity and commutativity of disjunction and conjunction, and De Morgan's laws. However, disjunction and conjunction are not idempotent, nor does a distributive law hold. This means that conceptors do not form a Boolean algebra. On the other hand, the subalgebra of hard conceptors do form a Boolean algebra [11].

2.3 Application of Conceptors to Japanese Vowel Classification

Jaeger [11] uses conceptors to learn and classify audio signals for Japanese vowels obtained from nine different speakers. Classification of a recording using a conceptor C_j (for speaker j) is done with the quadratic form[2]

$$x^T C_j x$$

where x is an observation of the reservoir state, while the reservoir is fed with the input signal.

Beyond this positive classification, using the logical operations for conceptors, one can also do a negative classification in the sense of "this speaker is not any of the other eight speakers". This negative classification of x is done using

$$x^T \neg (C_1 \vee \cdots \vee C_{j-1} \vee C_{j+1} \vee \cdots \vee C_n)x$$

---

[1] See `https://www.youtube.com/watch?v=DkS_Yw1ldD4`.
[2] Cf. the formula $z^T C_j^+ z$ on p.75 [11].

Jaeger used 12-channel recordings of short utterance of 9 male Japanese speakers. There are 270 training recordings and 370 test recordings. After forming conceptors using the training data, speaker recognition on the test data could be done with an error rate of less than 1%. Moreover, using the quasi-Boolean operations, an incremental extension of the model is possible, while avoiding catastrophic forgetting that occurs with other approaches.

2.4   The Subconceptor Relation, and Conceptor Logic

Conceptors can be partially ordered, providing a concept hierarchy, as it is known from ontologies and other knowledge representation formalisms. In [11], a partial order on conceptors is given by $C \leq D$ iff $D - C$ is positive semi-definite (this is the so-called Löwner ordering).

In [11], it is shown that $C \leq D$ iff there is some E with $C \vee E = D$ iff there is some conceptor E with $C = D \wedge E$. However, the usual definition of the partial order in a lattice via $C \leq D$ iff $C = C \wedge D$ does not work. This shows that the Löwner ordering $\leq$ does not form a lattice with the above operations.[3] Fig. 3 shows the Löwner ordering among all 2×2 conceptors that use multiples of $\frac{1}{4}$ for their entries. Note that for conception vectors, the ordering is just component-wise. Using some simple calculations on eigenvalues, we can characterise 2×2 conceptors as those matrices $\begin{pmatrix} a & b \\ c & d \end{pmatrix}$ with $a, d \in [0, 1]$, $b = c$ and

$$|b| \leq \sqrt{\min(ad, ad - a - d + 1)},$$

and the Löwner ordering as given by

$$\begin{pmatrix} a_1 & b_1 \\ b_1 & d_1 \end{pmatrix} \leq \begin{pmatrix} a_2 & b_2 \\ b_2 & d_2 \end{pmatrix} \text{ iff } a_1 \leq a_2, d_1 \leq d_2 \text{ and } |b_2 - b_1| \leq \sqrt{(a_2 - a_1)(d_2 - d_1)}.$$

While it is clear that the Löwner ordering does not form a lattice with the above Quasi-Boolean operations, in principle, there could be other operations turning it into a lattice. However, this is not the case, as our following example shows:

---

[3]Indeed, on all symmetric matrices, the Löwner ordering forms an anti-lattice: any two matrices have a supremum only if they are already comparable (and hence one of them is the supremum) [17]. However note that a lattice can be simultaneously an anti-lattice, e.g. in the case of a total order. For conceptors, this happens iff the dimension M is 1.

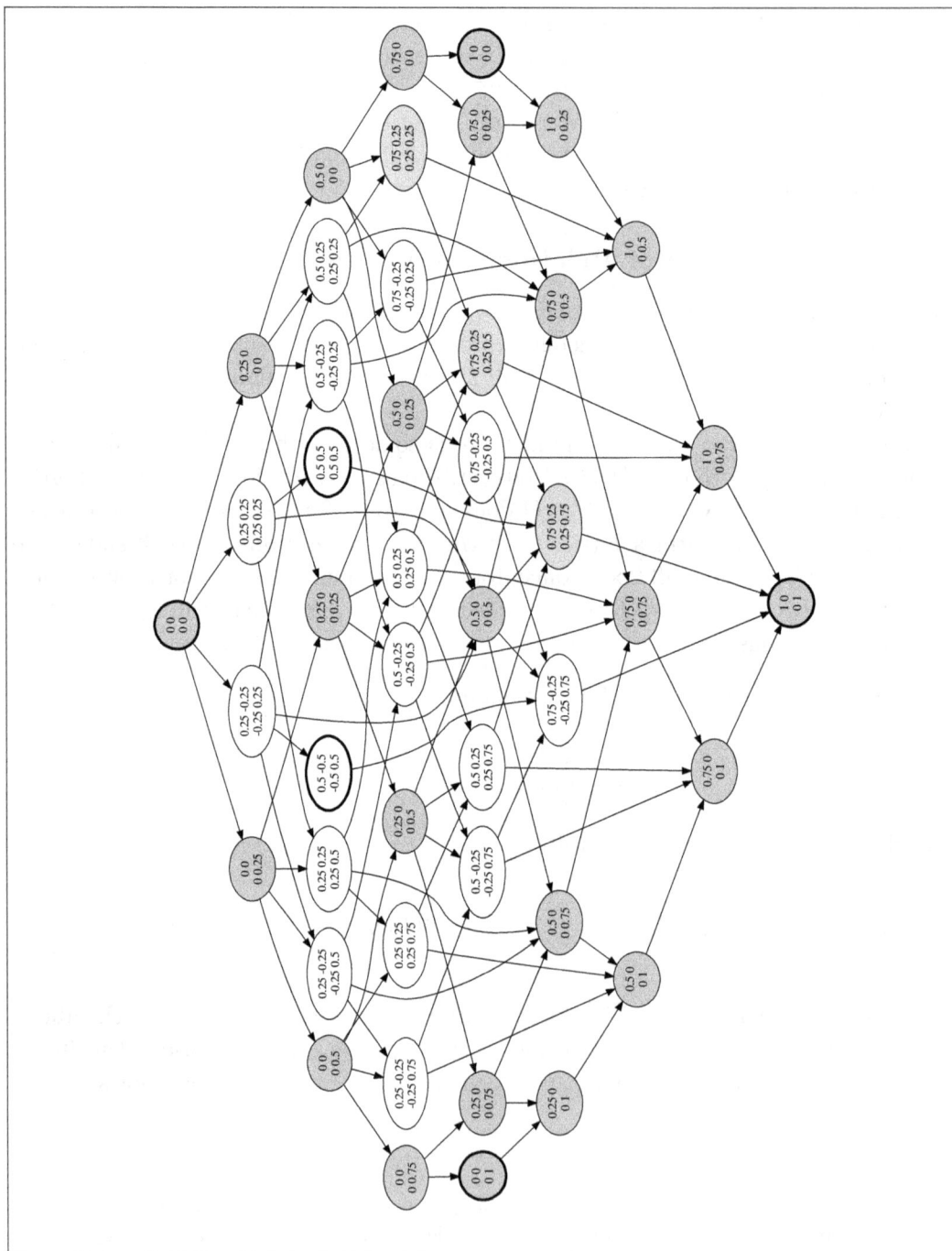

Figure 3: The Löwner ordering for some 2×2 conceptors. Conception vectors are show in grey, and hard conceptors with a thick borderline.

Example 1. *The Löwner ordering $\leq$ on conceptors is not a lattice, because we have a pair of conceptors with three incomparable common upper bounds:*

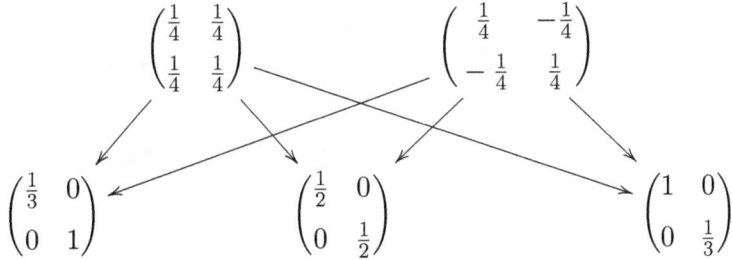

$$
\begin{pmatrix} \frac{1}{4} & \frac{1}{4} \\ \frac{1}{4} & \frac{1}{4} \end{pmatrix} \qquad\qquad \begin{pmatrix} \frac{1}{4} & -\frac{1}{4} \\ -\frac{1}{4} & \frac{1}{4} \end{pmatrix}
$$

$$
\begin{pmatrix} \frac{1}{3} & 0 \\ 0 & 1 \end{pmatrix} \qquad \begin{pmatrix} \frac{1}{2} & 0 \\ 0 & \frac{1}{2} \end{pmatrix} \qquad \begin{pmatrix} 1 & 0 \\ 0 & \frac{1}{3} \end{pmatrix}
$$

For reasoning about the subconceptor relation, Jaeger defines two logics for conceptors, an extrinsic and intrinsic one. Both logics feature conceptor terms using the above defined operations, and feature subconceptor relations (i.e. the Löwner ordering) between such conceptor terms as atomic formulas. We will briefly recall the extrinsic logic in simplified form[4], the intrinsic logic not being so relevant to our topic. Extrinsic conceptor logic builds a first-order logic on top of the subconceptor relation. This means that algebraic laws for the conceptor operations or properties of the Löwner ordering can be formally studied in this logic.

Definition 3 (Extrinsic conceptor logic [11]). *Extrinsic conceptor logic is parameterised over a dimension $M \in \mathbb{N}$. Signatures consist of a set of constants. Models interpret these as M-dimensional conceptors in $[0,1]^M$. Conceptor terms are formed from constants c from the signature, the constants 0 and 1, conceptor variables, quasi-Boolean operations, aperture adaption, and interpolation:*

$$
C ::= c \mid x \mid \bot \mid \top \mid \neg x \mid C_1 \vee C_2 \mid C_1 \wedge C_2 \mid \varphi(C, r) \mid \beta_b(C_1, C_2)
$$

The above definition of quasi-Boolean operations, aperture and interpolation for conceptors leads to an interpretation of terms in a model and a variable valuation. Atomic formulas $C_1 \leq C_2$ are inequalities between conceptor terms C_1, C_2. They are interpreted using the Löwner ordering. Complex formulas are formed from these using the standard first-order Boolean connectives and quantifiers:

$$
F ::= C_1 \leq C_2 \mid \neg F \mid F_1 \vee F_2 \mid F_1 \wedge F_2 \mid \forall x.F \mid \exists x.F
$$

Their interpretation follows that in standard first-order logic.

Note that extrinsic conceptor logic features two levels of (quasi-)Boolean connectives: one level inside conceptor terms, not satisfying the laws of Boolean algebra, and one level within formulas, satisfying the laws of Boolean algebra.

---

[4]For simplicity, we here omit signature morphisms.

3 A Fuzzy Logic for Conceptors

Both extrinsic and intrinsic conceptor logic have the drawback that they are based on crisp statements $C_1 \leq C_2$ that can be either true or false. However, conceptors (and conceptors) abstract, classify and reconstruct signals not in a precise and crisp manner, but are rather fuzzy in nature. Hence, the central thesis of this paper is:

> Conceptors and conceptors behave like fuzzy sets, and their logic should be a fuzzy logic.

3.1 A Generalised De Morgan Triplet

This thesis is backed up by the observation that conceptors satisfy the usual algebraic laws for fuzzy sets, namely that of a De Morgan triplet — here generalised from the laws for fuzzy truth values, as follows:

Definition 4 (strong negation [1]). *Let $(X, \leq)$ be a partial order with minimum 0 and maximum 1. A function $N : X \to X$ is called a generalised[5] strong negation, if*

- $N(0) = 1, N(1) = 0$

- $x < y$ *implies* $N(x) > N(y)$ *(strict anti-monotonicity)*

- $N(N(x)) = x$ *(involution)*

Definition 5 (t-norm, t-conorm [1]). *A function $T : X^2 \to X$ is called a generalised t-norm, if it satisfies the following properties:*

- *T1: $T(x, 1) = x$ (identity)*

- *T2: $T(x, y) = T(y, x)$ (commutativity)*

- *T3: $T(x, T(y, z)) = T(T(x, y), z)$ (associativity)*

- *T4: If $x \leq u$ and $y \leq v$ then $T(x, y) \leq T(u, v)$ (monotonicity)*

A function $S : X^2 \to X$ is called a generalised t-conorm, if it satisfies T2-T4 above (adapted to S) and

- *S1: $S(x, 0) = x$ (identity)*

---

[5]Generalised from $[0, 1]$ to an arbitrary partial order $(X, \leq)$.

Definition 6 (De Morgan triplet [1]). *A triplet $(S : X^2 \to X, T : X^2 \to X, N : X \to X)$ is called a generalised De Morgan triplet if T is a generalised t-norm, S is a generalised t-conorm, N is a generalised strong negation, such that De Morgan's law is satisfied:*

$$S(x,y) = N(T(N(x), N(y)))$$

Proposition 1. *conceptors and their quasi-Boolean operations form a generalised De Morgan triplet $(\vee, \wedge, \neg)$ when the minimum 0 is the zero matrix 0, and 1 the unit matrix I.*

Proof. Follows from proofs in [11] and straightforward calculations. $\square$

3.2 A Fuzzy Subconceptor Relation

The crisp subconceptor relation makes rather sharp distinctions. For example, $\begin{pmatrix} 0.9 & 0.001 \\ 0.001 & 0.9 \end{pmatrix} \not\preceq \begin{pmatrix} 0.9 & 0 \\ 0 & 0.9 \end{pmatrix}$, although both conceptors differ only minimally, and for classification purposes, there would be not much difference between them (of course, realistic classification examples like in the Japanese vowel example need much larger conceptors). We generally expect that a subconceptor relation should be immune to small disturbances and noise in signals. This can be achieved with a fuzzy subconceptor relation, which also would be in accordance with the finding that conceptors behave like fuzzy sets.

Therefore, an important question is how the Löwner ordering generalises to the fuzzy setting. The resulting truth value will then not be true or false, but will be a member of $[0, 1]$, the space of fuzzy truth values. Note that the crisp Löwner ordering is defined as $C \leq D$ iff $D - C$ is positive semi-definite, which in turn holds iff all eigenvalues of $D - C$ are non-negative, i.e. $eig(D - C) \subseteq \mathbb{R}_0^+$. Our fuzzy generalisation of the Löwner ordering is defined as:

Definition 7.

$$C \preceq D = 1 + mean(nullify\_positives(eig(D - C)))$$

where

$$nullify\_positives(x) = \begin{cases} 0 & x \geq 0 \\ x & otherwise \end{cases}$$

Note that $eig(D - C)$ is considered as multiset (that is, an eigenvalue occurring multiple times counts multiply in the mean).
Furthermore, fuzzy equivalence is defined as

$$C \cong D = \min(C \preceq D, D \preceq C)$$

Proposition 2. $C \preceq D \in [0,1]$, *hence this is indeed a fuzzy truth value.*

Proof. Since C and D are conceptors, all their eigenvalues are in $[0,1]$. By Weyl's inequality, for the minimal eigenvalues, we have $\lambda_{min}(D-C) \geq \lambda_{min}(D) + \lambda_{min}(-C) \geq 0 - 1 = -1$ and for the maximal ones $\lambda_{max}(D-C) \leq \lambda_{max}(D) + \lambda_{max}(-C) \leq 1 + 0 = 1$. Hence, the eigenvalues of $D - C$ are in $[-1,1]$. Therefore, $eig(D-C) \cap \mathbb{R}_0^- \subseteq [-1,0]$, and $1 + mean(nullify\_positives(eig(D-C))) \in [0,1]$. $\qquad\square$

The relation of this fuzzy ordering to the crisp one is as follows:

Proposition 3.
$$C \preceq D = 1 \text{ iff } C \leq D$$

Proof. $C \preceq D = 1$ iff $mean(nullify\_positives(eig(D-C))) = 0$ iff all eigenvalues of $D - C$ are non-negative iff $C \leq D$. $\qquad\square$

Example 2 (Examples for the fuzzy subconceptor relation). *We illustrate the fuzzy subconceptor relation using some simple 2×2 matrices:*

- $\begin{pmatrix} 0.9 & 0.1 \\ 0.1 & 0.9 \end{pmatrix} \not\preceq \begin{pmatrix} 0.9 & 0 \\ 0 & 0.9 \end{pmatrix}$

- $\begin{pmatrix} 0.9 & 0.1 \\ 0.1 & 0.9 \end{pmatrix} \preceq \begin{pmatrix} 0.9 & 0 \\ 0 & 0.9 \end{pmatrix}$ *is 0.9*

- $\begin{pmatrix} 1 & 0 \\ 0 & 1 \end{pmatrix} \not\preceq \begin{pmatrix} 0.9 & 0.1 \\ 0.1 & 0.9 \end{pmatrix}$

- $\begin{pmatrix} 1 & 0 \\ 0 & 1 \end{pmatrix} \preceq \begin{pmatrix} 0.9 & 0.1 \\ 0.1 & 0.9 \end{pmatrix}$ *is 0.8*

Example 3. *We can also set a threshold, e.g. 0.85, and ask whether conceptors relate fuzzily with a value above this threshold. With threshold 0.85, we have*

but

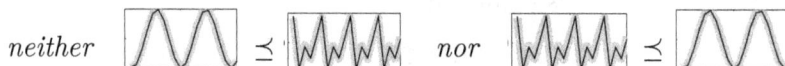

For the Japanese vowel speakers, we need a different threshold, e.g. 0.9. Then two speakers are equivalent iff they are identical. By contrast, with threshold 0.7, all Japanese vowel speakers are equivalent ($\cong$), showing that the sensor data allows no easy discrimination of the speakers.

Proposition 4. $C \preceq D = \neg D \preceq \neg C$

How close is a conceptor to the top element of the Löwner ordering, the unit matrix? Using the fuzzy subconceptor relation, the *weight* of a conceptor C can be defined as

$$w(C) = \top \preceq C$$

and by Prop. 4, this is equal to $\neg C \preceq \bot$.

Example 4. • $w(\ $$\ ) \approx 0.05$

• $w(\ $$\ ) \approx 0.2$

• $w(JP_j) \approx 0.2$ *for* $j = 1, \ldots, 9$

• $w(\neg(JP_1 \vee \cdots \vee JP_{j-1} \vee JP_{j+1} \vee \cdots \vee JP_n)) \approx 0.6$ *for* $j = 1, \ldots, 9$

Jaeger introduces the *quota* of a conceptor as the mean value of the singular values. The quota is a measure for the conceptor to learn new information (e.g. by taking disjunctions with other conceptors). A quota near 1 means that the capacity of the conceptor is nearly exceeded. See [11] for discussion and examples.

Proposition 5. *Weight and quota coincide.*

Proof. We have

$$
\begin{aligned}
w(C) &= \top \preceq C = I \preceq C \\
&= 1 + mean(nullify\_positives(eig(C - I))) \\
&\overset{*}{=} 1 + mean(eig(C - I)) \\
&\overset{**}{=} 1 + mean(eig(C)) - 1 \\
&= mean(eig(C)) \\
&\overset{***}{=} mean(svd(C)) \\
&= quota(C).
\end{aligned}
$$

Equation * holds because $I - C$ is a conceptor and thus has no negative eigenvalues. Hence, $C - I$ does not have positive eigenvalues. Equation ** can be seen as follows: $(C-I)v = (\lambda-1)v$ iff $((C-I)-(\lambda-1)I)v = 0$ iff $(C-\lambda I)v = 0$ iff $Cv = \lambda v$. Equation *** holds because for conceptors, eigenvalues and singular values coincide. □

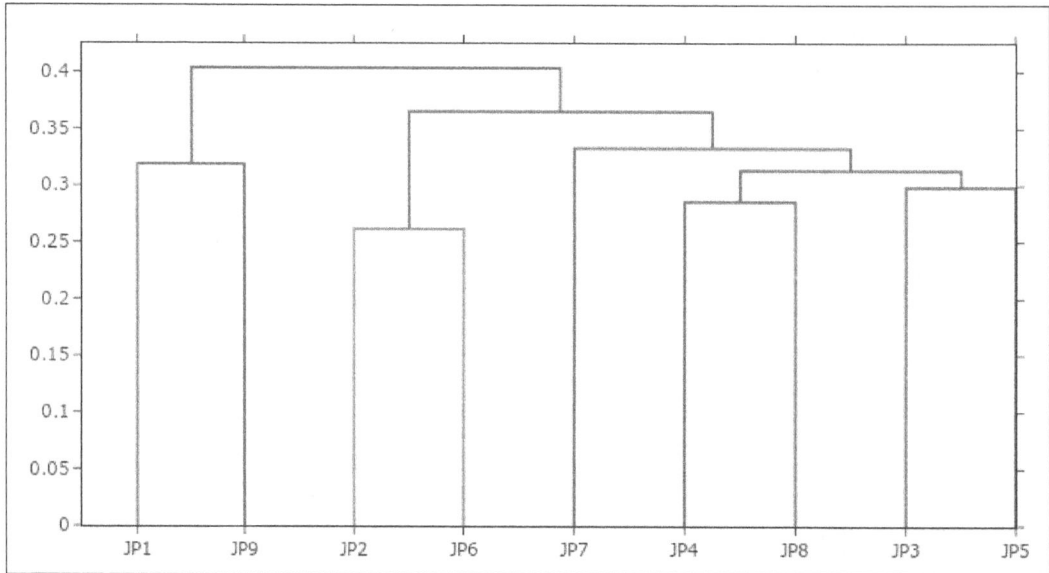

Figure 4: Dendrogram for Japanese speakers

This result is an indicator that the fuzzy conceptor relation $\preceq$ has a natural definition.

Coming back to the Japanese vowels example, we can use the fuzzy conceptor equivalence relation to create a dissimilarity matrix for the nine Japanese speakers. As distance between two speakers, we use negation (via $1-x$) of the fuzzy equivalence between the corresponding conceptors. When feeding this dissimilarity matrix into a simple hierarchical clustering algorithm [6], we obtain the dendrogram shown in Fig. 4.

Now the added value of using conceptors compared to mere clustering is that we can easily build conceptors from clusters, using disjunction (which can also be used in the agglomeration process when updating the dissimilarity matrix while merging clusters). For example, from the clusters in Fig. 4, we can choose to build e.g. three conceptors JP1 $\vee$ JP9, JP2 $\vee$ JP6 and JP3 $\vee$ JP4 $\vee$ JP5 $\vee$ JP7 $\vee$ JP8. These conceptors then will capture little "dialects" that can be used for further classification and clustering. The resulting concept hierarchy is crisp:

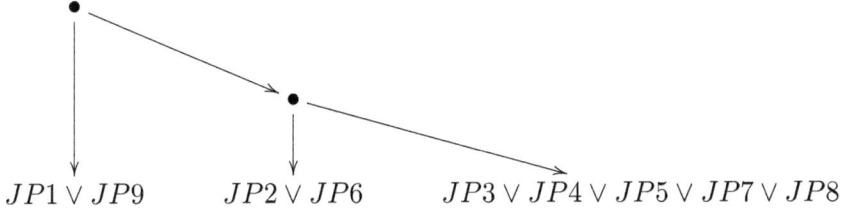

$$JP1 \vee JP9 \qquad JP2 \vee JP6 \qquad JP3 \vee JP4 \vee JP5 \vee JP7 \vee JP8$$

3.3 Fuzzy Conceptor Logic

In the previous section, we have defined distance between speakers as

$$1 - min(C \preceq D, D \preceq C),$$

which in terms of fuzzy logic can be rewritten as

$$\neg(C \preceq D \wedge D \preceq C).$$

That is, we have used propositional logic on top of the fuzzy subconceptor relation. We now extend this to fuzzy first-order logic. We also add fuzzy membership as atomic statements. This means that we can use fuzzy conceptor logic for reasoning about concept hierarchies formed by conceptors, but also about specific signals and their membership in these conceptors.

Definition 8 (Fuzzy conceptor logic). *Fuzzy conceptor logic is parameterised over a dimension $M \in \mathbb{N}$. The logic is two-sorted, distinguishing individuals and conceptors. Signatures consist of a set of constants for individuals and one for conceptors. Conceptor terms are formed from constants from the signature, the constants $\perp$ and $\top$, conceptor variables, quasi-Boolean operations, and aperture adaption:*

$$C ::= c \mid x \mid \perp \mid \top \mid \neg x \mid C_1 \vee C_2 \mid C_1 \wedge C_2 \mid \varphi(C, r) \mid \beta_b(C_1, C_2)$$

Complex formulas are formed from the following atomic formulas: (1) ordering relations $C_1 \preceq C_2$ between conceptor terms and (2) memberships $i \in C$ of individual constants in conceptors. Complex formulas may use the standard Boolean connectives and first-order quantifiers:

$$\begin{aligned} F \quad ::= \quad & i \in C \mid C_1 \preceq C_2 \mid \neg F \mid F_1 \vee F_2 \mid F_1 \wedge F_2 \mid F_1 \to F_2 \\ & \mid \forall x^i.F \mid \forall x^c.F \mid \exists x^i.F \mid \exists x^c.F \end{aligned}$$

Here, x^i denotes a variable ranging over individuals, and x^c one ranging over conceptors.

The semantics of fuzzy conceptor logic is as follows. Models interpret constants for individuals as vectors in $[0,1]^M$ and constants for conceptors as conceptor matrices in $[0,1]^{M \times M}$. Conceptor terms are evaluated as for extrinsic conceptor logic (i.e. using the operations on conceptors, see Def. 2).

The two types of atomic formulas have the following semantics, giving a fuzzy truth value in $[0,1]$:

$$\llbracket C_1 \preceq C_2 \rrbracket^M = \llbracket C_1 \rrbracket^M \preceq \llbracket C_2 \rrbracket^M$$
$$\llbracket i \in C \rrbracket^M = \tfrac{1}{M} (\llbracket i \rrbracket^M)^T \llbracket C \rrbracket^M \llbracket i \rrbracket^M$$

Note that the latter formula is a quadratic form. Quadratic forms have been used by Jaeger for classification of utterances of Japanese speakers (see the discussion at the end of Sect. 2) and also in [16]. We need to show that the quadratic form has range $[0,1]$. Since $\llbracket C \rrbracket^M$ is a positive semi-definite, we have that $(\llbracket i \rrbracket^M)^T \llbracket C \rrbracket^M \llbracket i \rrbracket^M \geq 0$. Moreover, we have $\tfrac{1}{M} (\llbracket i \rrbracket^M)^T \llbracket C \rrbracket^M \llbracket i \rrbracket^M \leq \tfrac{1}{M} \cdot \left\| \llbracket i \rrbracket^M \right\|^2 \cdot \lambda_{max}(\llbracket C \rrbracket^M) \leq \tfrac{1}{M} \sqrt{M}^2 \cdot 1 = 1$.

Interpretation of complex formulas follows that for fuzzy first-order logic. We need to interpret logical connectives and quantifiers as fuzzy operations on $[0,1]$ We use *min* as conjunction, *max* as disjunction and $1 - x$ as negation, and use infimum for universal quantification and supremum for existential quantification. Moreover, it is standard to define *residual implication* as:

$$R(x,y) = \sup\{t \in X \mid x \wedge t \leq y\}$$

Proposition 6. *For the conjunction $\wedge$ defined as min on $[0,1]$, residual implication amounts to*

$$R(x,y) = \begin{cases} 1 & x \leq y \\ y & otherwise \end{cases}$$

In particular, $x \leq y$ iff $R(x,y) = 1$.

Proof. If $x \leq y$, the inequation $min(x,t) \leq y$ even holds for $t = 1$. If not, $t = y$ is the largest value for which it holds. $\square$

Residual implication is illustrated in Fig. 5. Note that it is discontinuous at (x,x).

In fuzzy conceptor logic, we now can express dissimilarity between two conceptors C and D (used for clustering, see Fig. 4) as

$$\neg(C \preceq D \wedge D \preceq C)$$

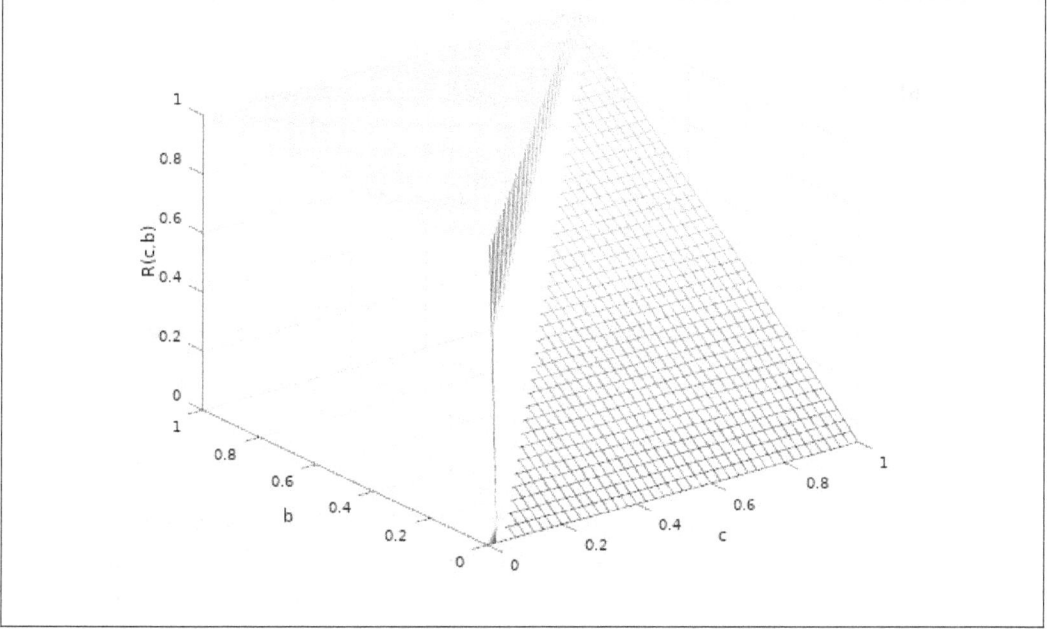

Figure 5: Residual implication $R(c, b)$.

Using the quantification capabilities of fuzzy conceptor logic, we can define two "subset" relations, namely 1) $C \preceq D$ and 2) the definition from classical set theory: $\forall x^i.(x^i \in C \rightarrow x^i \in D)$. We then have

Proposition 7. *If* $[\![C]\!]^M \leq [\![D]\!]^M$, *then* $[\![\forall x^i.(x^i \in C \rightarrow x^i \in D)]\!]^M = 1$.

Proof. If $[\![C]\!]^M \leq [\![D]\!]^M$, then by monotonicity of the scalar product, $[\![x^i \in C]\!]^M \leq [\![x^i \in D]\!]^M$. Hence, by Prop. 6, $[\![x^i \in C \rightarrow x^i \in D]\!]^M = 1$. $\qquad\square$

Example 5. *The converse does not hold: consider* $[\![C]\!]^M = \begin{pmatrix} 0.9 & -0.1 \\ -0.1 & 0.9 \end{pmatrix}$ *and*

$[\![D]\!]^M = \begin{pmatrix} 0.9 & 0 \\ 0 & 0.9 \end{pmatrix}$, *we have* $[\![C]\!]^M \not\leq [\![D]\!]^M$. *However, still* $[\![\forall x^i.(x^i \in C \rightarrow x^i \in D)]\!]^M = 1$. *This is because for* $x = [\![x^i]\!]^M$, $[\![x^i \in C]\!]^M = 0.9(x_1^2 + x_2^2) - 0.1x_1x_2$, *while* $[\![x \in D]\!]^M = 0.9(x_1^2 + x_2^2)$. *Hence* $[\![x^i \in C \rightarrow x^i \in D]\!]^M = 1$.

For conception vectors, we have a stronger property:

Proposition 8. *For conception vectors* $[\![C]\!]^M$, $[\![D]\!]^M$,

$$[\![\forall x^i.(x^i \in C \rightarrow x^i \in D)]\!]^M = \begin{cases} 1 & [\![C]\!]^M \leq [\![D]\!]^M \\ 0 & otherwise \end{cases}$$

741

Proof. One direction is Prop. 7. Concerning the other direction, if $[\![C]\!]^M \not\preceq [\![D]\!]^M$, then pick some $j \in \{1, \ldots, n\}$ with $[\![C]\!]_j^M > [\![D]\!]_j^M$. Let e_j be the j-th unit vector. Then for all $r \geq 0$, $r \cdot e_j^T [\![C]\!]^M r \cdot e_j > r \cdot e_j^T [\![D]\!]^M r \cdot e_j$, and $\lim_{r \to 0} R(r \cdot e_j^T [\![C]\!]^M r \cdot e_j, r \cdot e_j^T [\![D]\!]^M r \cdot e_j) = 0$. Therefore, $[\![\forall x^i.(x^i \in C \to x^i \in D)]\!]^M = 0$. $\qquad\square$

Corollary 1. *For conception vectors $[\![C]\!]^M$, $[\![D]\!]^M$,*

$$[\![[\forall x^i.(x^i \in C \to x^i \in D)] \to C \preceq D]\!]^M = 1$$

The corollary does not hold for conceptors in general, as Example 5 shows.

4 Conclusion

Our central thesis is that Jaeger's conceptor logic is best formalised as a fuzzy logic. We have shown that this fuzzy logic smoothly extends Jaeger's crisp logic. In particular, we have generalised his subconceptor relation (which is the Löwner ordering) to a fuzzy subconceptor relation. A potential application of this logic is the extended classification of speech signals. Based on Jaeger's organisation of vowel utterances by nine different Japanese speakers, we have used the fuzzy subconceptor relation to cluster these speakers. The advantage of our framework is that the can build conceptors from these clusters (using disjunction). The resulting conceptors correspond to little "dialects", and they can themselves directly be used for classification, but also for further clustering into taxonomies. Moreover, new (sensor) data can be integrated without re-training, again using conceptor disjunction. If several such taxonomies have to be merged, it is unlikely that conceptors match exactly, or a in the crisp subconceptor relation. Here, our fuzzy subconceptor relation could be useful for taxonomy (ontology) alignment [5].

In order to extend this approach to a framework for creating and maintaining ontologies based on sensor data, more research is needed. Conceptors are essentially modelling concepts (formally: unary predicates). For ontological modelling that extends mere taxonomies, an important question is: is it possible to model binary (or even n-ary) predicates in a similar way? If yes, then can we model quantifiers as projections, like in description logics? This could have advantages over the current formalisation of quantifiers, which behave in rather crisp way when interacting with membership, cf. Prop. 8. Moreover, currently, the domain of discourse for individuals consists of all vectors in $[0, 1]^M$. Would it be useful to restrict this to a subspace, such that e.g. vectors that would represent nonexistent objects can be excluded?

A peculiarity of the fuzzy conceptor logic that we have introduced is the difference between $x \in (C \vee D)$ and $x \in C \vee x \in D$. It is unclear how these two formulas relate, but it seems that the latter formula leads to a loss of information in some sense. A similar phenomenon arises with the formulas $\forall x^i.(x^i \in C \to x^i \in D)$ and $C \preceq D$. intuitively, they both express a form of fuzzy subconceptor relation, but further study is needed to understand the subtle differences.

Various algebraic and order-theoretic properties should be studied. We conjecture that $\preceq$ is a fuzzy partial order in the sense of [13].

Furthermore, future work should provide proof support for fuzzy conceptor logic and study the integration of inductive learning from examples and deductive reasoning. It should be straightforward to organise fuzzy conceptor logic as an L-institution in the sense of [3]. Then, for obtaining a proof calculus, the very general results of [3] can be applied, because our space of truth value forms a residuated lattice.

Finally, it will be crucial to study more application scenarios of conceptors in the rich field of time series. For example, we are currently working with weather data and energy data for modelling renewable energies.

Acknowledgements   We thank (in alphabetical order) Xu "Owen" He, Herbert Jaeger, Ulf Krumnack, Kai-Uwe Kühnberger, Fabian Neuhaus, Nico Potyka and Georg Schroeter for fruitful discussions.

References

[1] Barnabas Bede. *Mathematics of Fuzzy Sets and Fuzzy Logic*. Springer, 2013.

[2] Tarek R. Besold, Artur S. d'Avila Garcez, Sebastian Bader, Howard Bowman, Pedro M. Domingos, Pascal Hitzler, Kai-Uwe Kühnberger, Luís C. Lamb, Daniel Lowd, Priscila Machado Vieira Lima, Leo de Penning, Gadi Pinkas, Hoifung Poon, and Gerson Zaverucha. Neural-symbolic learning and reasoning: A survey and interpretation. *CoRR*, abs/1711.03902, 2017.

[3] Razvan Diaconescu. Graded consequence: an institution theoretic study. *Soft Comput.*, 18(7):1247–1267, 2014.

[4] Ivan Donadello, Luciano Serafini, and Artur S. d'Avila Garcez. Logic tensor networks for semantic image interpretation. In Carles Sierra, editor, *Proceedings of the Twenty-Sixth International Joint Conference on Artificial Intelligence, IJCAI 2017, Melbourne, Australia, August 19-25, 2017*, pages 1596–1602. ijcai.org, 2017.

[5] Jérôme Euzenat and Pavel Shvaiko. *Ontology Matching, Second Edition*. Springer, 2013.

[6] Brian S. Everitt, Sabine Landau, Morven Leese, and Daniel Stahl. *Cluster Analysis*. Wiley, 2011. 5th Edition.

[7] M.M. Gupta and J. Qi. Theory of T-norms and fuzzy inference methods. *Fuzzy Sets and Systems*, 40:431–450, 1991.

[8] Xu He and Herbert Jaeger. Overcoming catastrophic interference using conceptor-aided backpropagation. In *Sixth International Conference on Learning Representations, ICLR 2018*, 2018.

[9] Eyke Hüllermeier. Fuzzy sets in machine learning and data mining. *Appl. Soft Comput.*, 11(2):1493–1505, 2011.

[10] Herbert Jaeger. Conceptors: an easy introduction. *CoRR*, abs/1406.2671, 2014.

[11] Herbert Jaeger. Controlling recurrent neural networks by conceptors. *CoRR*, abs/1403.3369, 2014.

[12] Herbert Jaeger. Using conceptors to manage neural long-term memories for temporal patterns. *Journal of Machine Learning Research*, 18:13:1–13:43, 2017.

[13] Hongliang Lai and Dexue Zhang. Fuzzy preorder and fuzzy topology. *Fuzzy Sets and Systems*, 157(14):1865–1885, 2006.

[14] Yann LeCun, Yoshua Bengio, and Geoffrey E. Hinton. Deep learning. *Nature*, 521(7553):436–444, 2015.

[15] E. Smith and S. Kosslyn. *Cognitive psychology: Mind and brain*. Upper Saddle River, NJ: Prentice-Hall Inc., 2006.

[16] Richard Socher, Danqi Chen, Christopher D. Manning, and Andrew Y. Ng. Reasoning with neural tensor networks for knowledge base completion. In Christopher J. C. Burges, Léon Bottou, Zoubin Ghahramani, and Kilian Q. Weinberger, editors, *Advances in Neural Information Processing Systems 26: 27th Annual Conference on Neural Information Processing Systems 2013. Proceedings of a meeting held December 5-8, 2013, Lake Tahoe, Nevada, United States.*, pages 926–934, 2013.

[17] Nikolas Stott. Maximal lower bounds in the Loewner order. To appear in Proceedings of the AMS, 2017. Also https://arxiv.org/abs/1612.05664.

[18] Lotfi A. Zadeh. Fuzzy sets. *Information and Control*, 8(3):338–353, 1965.

Received 18 June 2018

MODELLING IDENTITY RULES WITH NEURAL NETWORKS

TILLMAN WEYDE

Research Centre for Machine Learning, Department of Computer Science, City, University of London, United Kingdom
t.e.weyde@city.ac.uk

RADHA MANISHA KOPPARTI*

Research Centre for Machine Learning, Department of Computer Science, City, University of London, United Kingdom
radha.kopparti@city.ac.uk

Abstract

In this paper, we show that standard feed-forward and recurrent neural networks fail to learn abstract patterns based on identity rules. We propose Relation Based Pattern (RBP) extensions to neural network structures that solve this problem and answer, as well as raise, questions about integrating structures for inductive bias into neural networks.

Examples of abstract patterns are the sequence patterns ABA and ABB where A or B can be any object. These were introduced by Marcus et al (1999) who also found that 7 month old infants recognise these patterns in sequences that use an unfamiliar vocabulary while simple recurrent neural networks do not. This result has been contested in the literature but it is confirmed by our experiments. We also show that the inability to generalise extends to different, previously untested, settings.

We propose a new approach to modify standard neural network architectures, called Relation Based Patterns (RBP) with different variants for classification and prediction. Our experiments show that neural networks with the appropriate RBP structure achieve perfect classification and prediction performance on synthetic data, including mixed concrete and abstract patterns. RBP also improves neural network performance in experiments with real-world sequence prediction tasks.

We discuss these finding in terms of challenges for neural network models and identify consequences from this result in terms of developing inductive biases for neural network learning.

We would like to thank the anonymous reviewers of this article for their valuable comments and suggestions that helped to improve this article.

*Funded by a PhD studentship from City, University of London

1   Introduction

Despite the impressive development of deep neural networks over recent years, there has been an increasing awareness that there are some tasks that still elude neural network learning or need unrealistic amounts of data. Humans, on the other hand, are remarkably quick at learning and abstracting from very few examples. Marcus [1] showed in an experiment that 7-month old infants already recognise sequences by identity rules, i.e. which elements are repeated, after just two minutes of familiarization. In that study a simple recurrent neural network model was also tested and it failed to generalise these identity rules to new data.

In this study, we re-visit this problem and evaluate the performance of frequently used standard neural network models in learning identity rules. More specifically, we find that feed-forward and recurrent neural networks (RNN) and their gated variants (LSTM and GRU) in standard set-ups clearly fail to learn general identity rules presented as classification and prediction tasks.

We tackle this problem by proposing *Relation Based Patterns* (RBP), which model identity relationships explicitly as extensions to neural networks for classification and prediction. We show experimentally that on synthetic data the networks with suitable RBP structures learn the relevant rules and generalise with perfect classification and prediction. We also show that this perfect performance extends to mixed rule-based and concrete patterns, and that RBP improves prediction on real-world language and music data.

Identity rules are clearly in the hypothesis space of the neural networks, but the networks fail to learn them by gradient descent. We identify that both the comparison of related input neurons and of input tokens needs to be predefined in the network to learn general rules from data. The RBP structures introduce this inductive bias in the neural networks and thus enable the learning of identity rules by standard neural networks.

Our contributions in this paper are specifically:

- we evaluate several common NN architectures: feed-forward networks, RNN, GRU, and LSTM, in novel settings, and find that they fail to learn general identity rules;

- we identify reasons that prevent the learning process from being successful in this context;

- we propose the Relation Based Patterns, a new method to enable the learning of identity rules within the regular network structure;

- we show in experiments that identity rules can be learnt with RBP structure on artificial data, including mixed rule-based and concrete patterns, and that they improve performance in real-world prediction tasks;

The remainder of this paper is structured as follows. Section 2 introduces related work on modelling identity rules. Section 3 presents results of our experiments with standard neural network architectures. Section 4 presents our RBP model and its different variants. Section 5 presents the results of experiments using RBP structures. Section 6 addresses the application of RBP to mixed patterns and real data. Section 7 discusses the implications of the presented experimental results and Section 8 concludes this paper.

2 Related work

Our task is the learning of rules from sequential data. This is often seen as grammar learning, on which there have been many studies in psychology. [2] made an early contribution on implicit learning and generalisation. Subsequently, [3, 4] studied specifically the knowledge acquired during artificial grammar learning tasks.]

The specific problem we are addressing in this study is the recognition of abstract patterns that are defined by the identity relation between tokens in a sequence. In the well-known experiments by [1], infants were exposed to sequences of one of the forms *ABA* or *ABB*, e.g. 'la di la' or 'la di di', for a few minutes in the familiarisation phase.

In the test phase the infants were exposed to sequences with a different vocabulary (e.g. 'ba tu ba' and 'ba tu tu') and they showed significantly different behaviour depending on whether the sequences exhibited the form they were familiarised with or not.

This type of pattern only depends on the equality between elements of the sequence and after successful learning it should be recognisable independently of the vocabulary used. However, [1] also showed that simple recurrent Elman networks were not able to perform this learning task. This finding sparked an exchange about whether human speech acquisition is based on rules or statistics and the proposal of several neural networks models that claimed to match the experimental results. [5] and [6] proposed a solution based on a distributed representation of the input and on pre-training where the network is first trained to recognise repeated items in a sequence. The network is subsequently trained on classifying ABA vs ABB patterns. Only [6] reports specific results and has only 4 test data points, but 100% accuracy. However, [7] reported that they could not recreate these results.

[8] and [9] suggested solutions which are based on modified network architectures and training methods. [10] could not replicate the results by [9] and found that the models by [8] do not generalise. The claims by [10] were again contested by [11]. A number of other methods were suggested that used specifically designed network architectures, data representations, and training methods, such as [12, 13, 14, 15]. More recent work by [16] suggests that prior experience or pre-defined context representation ("pre-training" or "pre-wiring") is necessary for the network to learn general identity rules when using echo state networks. While these works are interesting and relevant, they do not answer our question whether more commonly used network architectures can learn general identity rules.

The discussion of this problem is part of a wider debate on the systematicity of language learning models, which started in the 1980s and 1990s [17, 18]. This debate, like the more specific one on identity rules, has been characterised by claims and counter-claims [19, 20, 21, 22, 23, 24], which, as stated by [25], often suffer from a lack of empirical grounding. Very recently, the work in [26] has defined a test of systematicity in a framework of translation, applied it to standard *seq2seq* neural network models [27]. They found that generalisation occurs in this setting, but it depends largely on the amount and type of data shown, and does not exhibit the extraction and systematic application of rules in the way a human learner would.

In most of the studies above, the evaluation has mostly been conducted by testing whether the output of the network shows a statistically significant difference between inputs that conform to a trained abstract pattern and those that do not. From a machine learning perspective, this criterion is not satisfactory as we, like [26], would expect that an identity rule should always be applied correctly once if it has been learned from examples, at least in cases of noise-free synthetic data. We are therefore interested in the question whether and how this general rule learning can be achieved with common neural network types for sequence classification.

This question also relates to recent discussions sparked by [28] about deep neural networks' need for very large amounts of training data, lack of robustness and lack of transparency as also expressed, e.g., by [29, 30, 31]. We surmise that these issues relate to the lack of generalisation beyond the space covered by the input data, i.e. extrapolation, which is generally seen as requiring an inductive bias in the learning system, but there is no general agreement about the nature or implementation of inductive biases for neural networks, e.g. [32, 33]. In recent years, there was a trend to remove human designed features from neural networks, and leave everything to be learned from the data [34]. We follow here the inverse approach, to add a designed internal representation, as we find that for the given problem standard neural network methods consistently fail to learn any suitable internal representation from the data.

3 Experiment 1: standard neural networks

We test different network architectures to evaluate if and to what extent recurrent and feed-forward neural networks can learn and generalise abstract patterns based on identity rules.

3.1 Supervised learning of identity rules

The problem in the experiment by [1] is an unsupervised learning task, as the infants in the experiments were not given instructions or incentives. However, most common neural network architectures are designed for supervised learning and there are also natural formulations of abstract pattern recognition as supervised learning task in the form of classification or prediction.

In our case, abstract patterns are defined by identity relations. Expressed in logic, they can be described using the binary equality predicate $eq(\cdot, \cdot)$. For a sequence of three tokens α, β, γ the rule-based patterns ABA and ABB can be described by the following rules:

$$ABA : \neg eq(\alpha, \beta) \wedge eq(\alpha, \gamma) \tag{1}$$

$$ABB : \neg eq(\alpha, \beta) \wedge eq(\beta, \gamma). \tag{2}$$

These rules are independent of the actual values of α, β, and γ and also called abstract patterns. Concrete patterns, on the other hand, are defined in terms of values of from a vocabulary $a, b, c, \dots$. E.g., sequences $a * *$, i.e. beginning with 'a', or $* bc$, ending with 'bc', can be formulated in logic as follows:

$$a * * : \alpha = \text{'}a\text{'} \tag{3}$$

$$* bc : \beta = \text{'}b\text{'} \wedge \gamma = \text{'}c\text{'}. \tag{4}$$

For the remainder of this article we use the informal notations ABA and $a * *$ as far as they are unambiguous in their context.

For classification, the task is to assign a sequence to a class, i.e. ABA or ABB, after learning from labelled examples. For prediction, the task is to predict the next token given a sequence of two tokens after exposure to sequences of one of the classes (e.g. only ABA, or ABB respectively). These tasks are suitable for the most commonly used neural network architectures.

3.2 Experimental set-up

Network set-up We use the Feed-forward Neural Network (FFNN) (also called Multi-layer Perceptron) [35], the Simple Recurrent Neural Network (RNN, also

749

called Elman network [36]), the Gated Recurrent Unit (GRU) network [37], and the Long Short Term Memory (LSTM) network [38]. For Prediction we only use the RNN and its gated variants GRU and LSTM.

The input to the networks is a one-hot encoded vector representing each token with n neurons, where n is the size of the vocabulary. In the case of the FFNN, we encode the whole sequence of 3 tokens as a vector of size $3n$. For the recurrent models, we present the tokens sequentially, each as an n-dimensional vector. We set the number of neurons in each hidden layer to (10, 20, 30, 40, 50), using 1 or 2 hidden layers. We use Rectified Linear Units (ReLUs) for the hidden layers in all networks. The output layer uses the softmax activation function. The number of output units is 2 for classification and the size of the vocabulary for prediction. We train with the Adam optimisation method [39], using initial learning rates of $0.01, 0.1, 0.2, 0.4$, and train with the synthetic datasets in one batch. We use regularisation with Dropout rates of 0.1, 0,2, 0.4 and set the number of epochs to 10, within which all trainings converged.

We conduct a full grid search over all hyperparameters using four-fold cross-validation to optimise the hyperparameters and determine test results. We run a total of 10 simulations for each evaluation and average the results. All experiments have been programmed in PyTorch and the code is publicly available.[1]

Datasets For performing the rule learning experiments, we artificially generate data in the form of triples for each of the experiments. We consider our sample vocabulary as $a...l$ (12 letters) for both prediction and classification tasks. We generate triples in all five abstract patterns: AAA, AAB, ABA, ABB, and ABC for the experiments. The sequences are then divided differently for the different cases of classification. For all the experiments we use separate train, validation, and test sets with 50%, 25%, and 25% of the data, respectively. All sampling (train/test/validation split, downsampling) is done per simulation.

3.3 Classification

First we test three different classification tasks as listed below. We use half the vocabulary for training and the other half for testing and validation (randomly sampled). We divide the sequences into two classes as follows, always maintaining an equal size of both classes:

1) ABA/ABB vs other: In task a) class one contains only pattern ABA while the other contains all other possible patterns (AAA, AAB, ABB, ABC) downsam-

---

[1]https://github.com/radhamanisha1/RBP-architecture

pled per pattern for class balance. The task is to detect whether $eq(\alpha, \gamma) \land \neg eq(\alpha, \beta)$ is true or false. Analogously, the task in b) ABB vs other is to detect $eq(\beta, \gamma) \land \neg eq(\alpha, \beta)$. This case corresponds to the experiment in [1], where only one rule-based pattern type is used for familiarisation.

2) ABA vs ABB: This task is like task 1 above, but only pattern ABB occurs in the second class, so that this task has less variance in the second class. We expected this task to be easier to learn because two equality predicates $eq(\alpha, \gamma), eq(\beta, \gamma)$ change their values between the classes and are each sufficient to indicate the class.

3) ABC vs other: In this case, class one (ABC) has no pair of equal tokens, while the *other* class has at least one of $eq(\alpha, \beta), eq(\alpha, \gamma), eq(\beta, \gamma)$ as *true*, i.e. detecting equalities without localising them is sufficient for correct classification.

In our experiments, the training converged quickly in all cases and the classification accuracy on the training data was 100%. The results on the test set are shown in Table 1. In all cases the baseline, corresponding to random guessing is 50%. This baseline is only exceeded for task 1) by the RNNs and their gated variants, and even then the accuracy is far from perfect at 55%.

Classification task	FFNN	RNN	GRU	LSTM
1a) ABA/other	50%	55%	55%	55%
1b) ABB/other	50%	55%	55%	55%
2) ABA/ABB	50%	50%	50%	50%
3) ABC/other	50%	50%	50%	50%

Table 1: Three classification tasks based on abstract patterns over 10 simulations. The numbers show test set accuracy after a grid search and cross validation as described in section 3.2. All values are rounded to the next percentage point.

3.4 Prediction

We performed prediction experiments on two tasks. In task 1) we train and test on ABA patterns and in task 2) on ABB. Training and test/validation set use different vocabularies. The training converged quickly in less than 10 epochs, and after training the classification accuracy on the training set is 100%.

The results on the test set are shown in Table 2. The baseline is $8.3\ldots\%$ as we have a vocabulary size of 12. We use again half the vocabulary (6 values) for training and half for validation/testing. The results show that the tested networks fail completely to make correct predictions. They perform below the baseline at 0% accuracy, which is mostly because they predict only tokens that appear in the training set but not in the test set.

Prediction task	RNN	GRU	LSTM
1) ABA	0%	0%	0%
2) ABB	0%	0%	0%

Table 2: Prediction results for two different abstract patterns. The numbers show test set prediction accuracy after a grid search and cross validation as described in section 3.2.

3.5 Discussion

The results show clearly that FFNNs, RNNs, GRUs and LSTMs do not learn general abstract patterns based on identity rules. This agrees with the previously reported experiments by [1]. However, since there was some conflicting evidence in the literature, the clarity of the outcome was not expected.

Questions raised This result raises the question of why these neural networks do not learn to generalise abstract patterns from data. There are two aspects worth considering for an explanation: the capacity of the network and the necessary information for the network to solve the problem.

Regarding the capacity: the solution to the task is in the hypothesis space of the neural networks, since proofs exist of universal approximation properties for feed-forward networks with unbounded activation functions [40] and of Turing-completeness for recurrent networks [41]. We will present a constructive solution below, putting that result into practice, with a design of network instances that solve the problem.

The relevant question, as has been pointed out by [16], is therefore why learning with backpropagation does not lead to effective generalisation here. There are three different steps that are necessary to detect identity rules: a comparison of input neurons, a comparison of tokens, represented by multiple neurons, and a mapping of comparison results to classes or predictions.

Vocabulary hypothesis A possible reason for the failure of the networks to generalise what we call the vocabulary hypothesis. It is based on the separated vocabulary in one-hot encoded representation. This leads to some input neurons only being activated in the training set and some only in the validation and test sets.

In order to learn suitable weights for an input comparison, there would have to be a suitable gradient of the weights of the outgoing connections from these inputs. If parts of the vocabulary do not appear in the training data, i.e. the activation of the corresponding input neurons is always zero during training, the weights of their outgoing connections will not be adapted. We therefore expect that the separation of the vocabulary prevents generalisation from the training to the test set as the weights going out from neurons that are used during testing have not been adapted by the gradient descent. Based on this consideration we conducted another experiment with a shared vocabulary.

This experiment is called ABA-BAB vs other. We again represent our vocabulary as $a...l$ (12 letters) for this task with train/validation/test split as 50%/25%/25%. Now we use the same vocabulary for training, validation, and testing, but we separate different sequences of the form ABA that use the same tokens between the training and validation/test sets. E.g., if ded is in the test set, then ede is in the training or validation set, so that there is no overlap in terms of actual sequences. Like in classification experiment 1), training converged quickly and resulted in perfect classification performance on the training set.

Classification task	FFNN	RNN	GRU	LSTM
ABA-BAB vs other	50%	50%	50%	50%

Table 3: Classification results on test sets with the same vocabulary used in test, validation and training set.

The results on the test set presented in Table 3 show performance at the baseline with no evidence of generalisation. This shows that activating all inputs by using a shared vocabulary is not sufficient to enable generalisation in the learning process.

Other explanations A second potential problem is which neurons should be compared. The FFNN has no prior information about neurons belonging to the same or different tokens or about which input neurons correspond to the same token values. In the RNN, one token is presented per time step, so that a comparison between the previous hidden state and the current input is possible as the same neurons are activated. However, with a full set of connections between the previous hidden layer and the current, there is no reason that relations between the same neurons at different time steps would be processed differently from different neurons.

On the other hand, if we had a representation that includes the information of which tokens are identical or different, then we would have all the information we need for a mapping, as these are the relations in which our defining rules are formulated (e.g. ABA is defined as $eq(\alpha, \gamma) \wedge \neg eq(\alpha, \beta)$). This idea has led to the Relation Based Pattern (RBP) model that we introduce in the next section and then evaluate with respect to its effect on both abstract and concrete pattern learning.

4 Design of Relation Based Pattern models

To address the inability of neural networks to generalise rules in neural network learning, we developed the Relation Based Pattern (RBP) model as a constructive solution, where the comparisons between input neurons and between tokens and the mappings to outputs are added as a predefined structure to the network. The purpose of this structure is to enable standard neural networks to learn abstract patterns based on the identity rules over tokens while retaining other learning abilities.

In the RBP model there are two major steps. The first step is defining comparison units for detecting identity relations, called DR units, and the second step is adding the DR units to the neural network.

4.1 Comparison units

Comparing neurons We assume, as before, that input is a one-hot encoded vector of the current token along with the $n-1$ previous vectors for a given context length n (in this study context length 3 for classification and 2 for prediction). We use comparison units, called DR units (differentiator-rectifier). As the name suggests, they apply a full wave rectification to the difference between two inputs: $f(x, y) = |x - y|$. The first level of DR units are DR_n units that are applied to every pair of corresponding input neurons (representing the same value) within a token representation, as shown in Figure 1.

Comparing tokens The next level of DR units are the DR_p units that sum the activations of the DR_n values that belong to one pair of tokens. Based on the sequence length n and vocabulary size a we create $k = a \times n(n-1)/2$ DR_n units for all the possible pairs of tokens and i.e. in our classification example, we have a sequence of 3 tokens and a vocabulary size of 12, i.e. $12 \times 3(3-1)/2 = 36 \times 3$ DR_n units. All the DR_n units for a pair of tokens are then summed in a DR_p unit using connections with a fixed weight of $+1$. E.g. we have $5 \times (5-1)/2 = 10$ DR_p units for a context of length 5. Figure 2a shows the network structure with DR_n and DR_p units.

For the prediction case, we also use the same approach to represent the difference between each input token and the next token (i.e., the target network output). We create n $DR_p out$ units that calculate the difference between each input in the

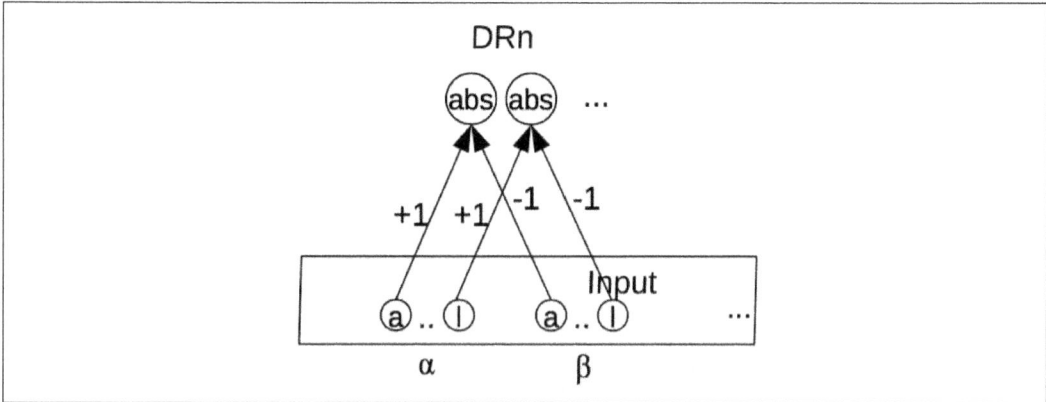

Figure 1: DR_n units comparing related inputs with an absolute of difference activation function. In one-hot encoding, there are k DR_n units for every pair of input tokens, where k is the vocabulary size.

given context and the next token. There are $k \times n$ $DR_n out$ units that compare the corresponding neurons for each pair of input/output tokens, in the same way as for the pairs of input tokens. The overall network structure is shown in Figure 2b.

4.2 Neural network integration

We combine the DR units (DR_n and DR_p) with the neural network models in early, mid and late fusion approaches we call RBP1, RBP2 and RBP3, as outlined below. The weights that connect DR_n units to input and output, and the DR_n to DR_p units and the offset layer are fixed, all other weights that appear in the following models are trainable with backpropagation.

Early Fusion (RBP1n/p) In this approach, DR_n or DR_p units are added as additional inputs to the network, concatenated with the normal input. In Figure 3, the RBP1n/p structure is depicted. We use early fusion in both the prediction and classification tasks.

Mid Fusion (RBP2) The DR_p units are added to the hidden layer. Figure 4a shows the mid fusion structure for the feed-forward network and Figure 4b for the recurrent network respectively. The RBP2 approach is used for classification and prediction tasks.

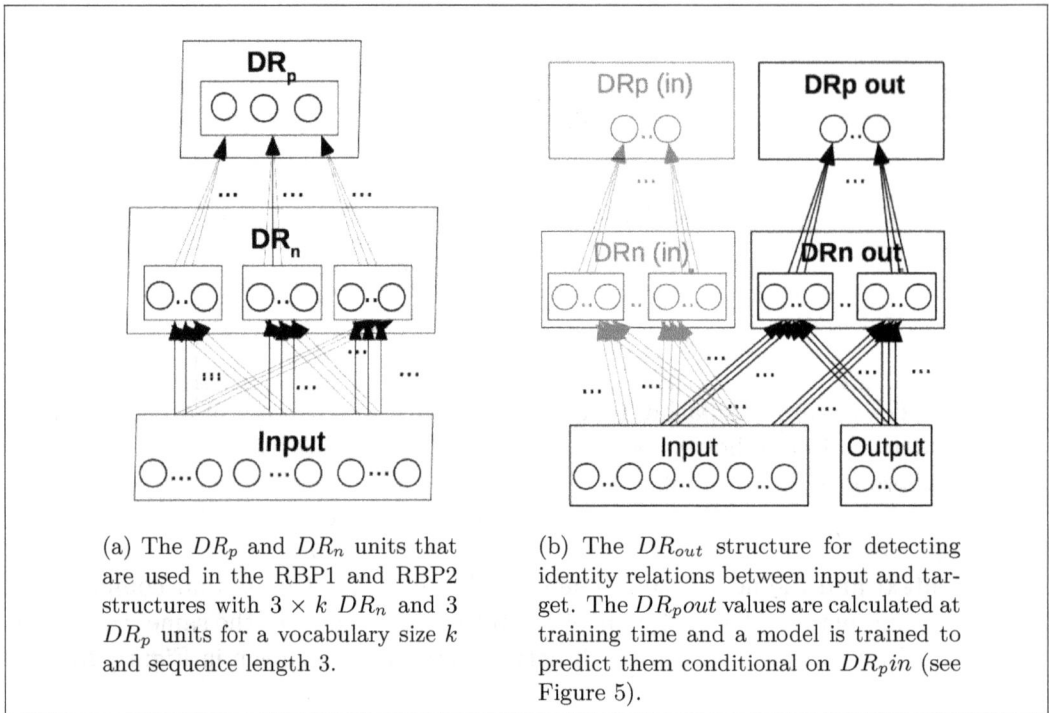

(a) The DR_p and DR_n units that are used in the RBP1 and RBP2 structures with $3 \times k$ DR_n and 3 DR_p units for a vocabulary size k and sequence length 3.

(b) The DR_{out} structure for detecting identity relations between input and target. The $DR_p out$ values are calculated at training time and a model is trained to predict them conditional on $DR_p in$ (see Figure 5).

Figure 2: DR_n and DR_p units for inputs (all RBP) and outputs (RBP3).

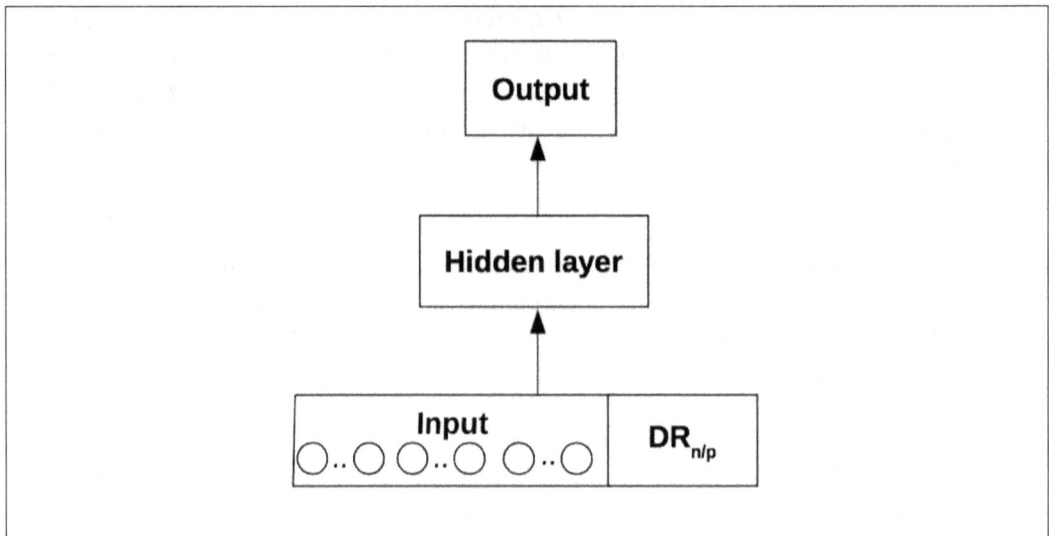

Figure 3: Overview of the RBP1n/RBP1p structure.

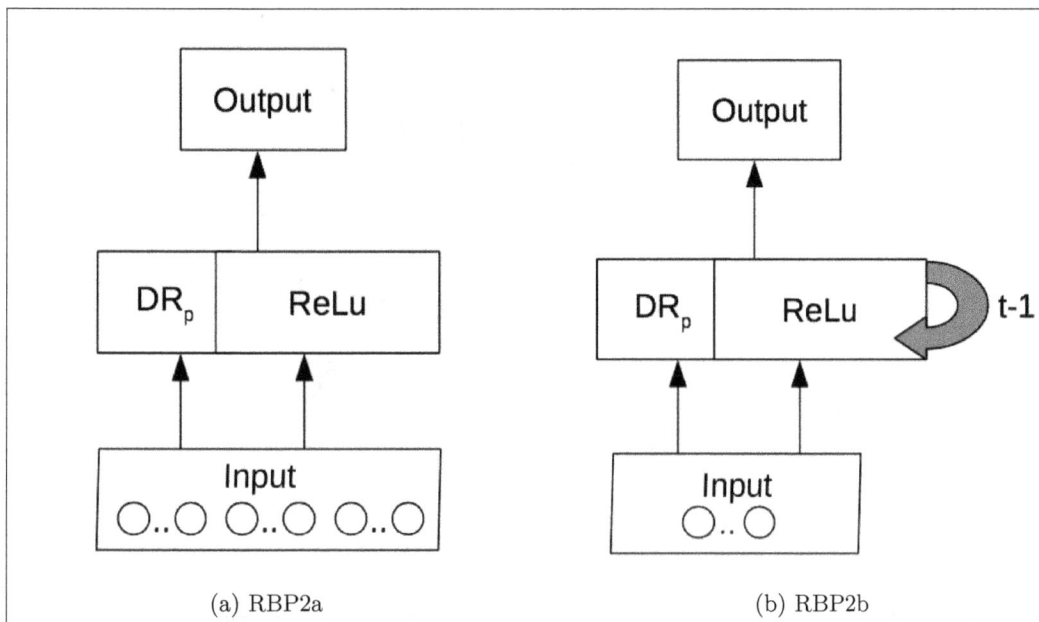

(a) RBP2a   (b) RBP2b

Figure 4: Overview of RBP2 approaches, where the $DR_p out$ units are concatenated to the hidden layer.

Late Fusion (RBP3) In this approach, we use the same structure as in RBP2 (we call it $DR_n in$ and $DR_p in$ in this context), and in addition we estimate the probability of identity relations between the input and the output, i.e., that the token in the current context is repeated as the next token. We use a structure called $DR_p out$ for this, and from there we project back to the vocabulary, to generate a probability offset for the tokens appearing in the context.

Figure 5 gives an overview of the RBP3 late fusion scheme. The $DR_p in$ units detect identities between the input tokens in the current context as before. The $DR_p out$ units model the identities between the context and the next token, as shown in the Figure 4b, where a repetition is encoded as 1, and a non-repeated token as a -1. During training we use teacher-forcing, i.e., we set the values of the $DR_p out$ units to the true values. We use a feed-forward neural network with one hidden layer to learn a mapping from the DR_{in} to the DR_{out}. This gives us an estimate of the DR_{out} units given the DR_{in} units. The DR_{out} values are then normalised subtracting the mean, and then mapped back to the output space (the one-hot vocabulary representation), using a zero value for the output values that don't appear in the input. These output offsets are then combined in a weighted sum (mixture of experts) with the output distribution estimated by the standard neural

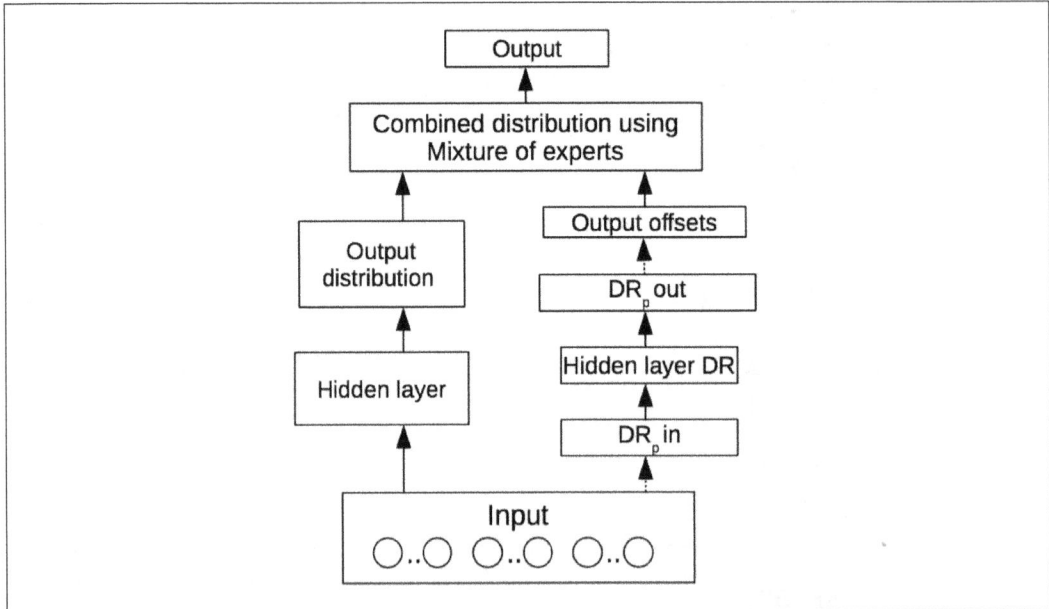

Figure 5: Overview of the RBP3 approach. The DR_pin values are calculated as in RBP2. From there, we use a fully connected layer to predict DR_pout (trained with teacher-forcing). The predicted DR_pout values are mapped back to the vocabulary (based on the context tokens) and used as probability offsets in a mixture of experts with the standard neural network in the left part of the diagram. All connections are trainable except *Input* to DR_pin and DR_pout to *Output offsets* (dotted arrows).

network (on the left side in Figure 5). The weights in the mixture are trainable. The outputs from the combined distribution of mixture of experts are clipped between [0,1] and renormalised. The final output distribution is a softmax layer providing the probability distribution over the vocabulary for the next token.

5 Experiment 2: neural networks with RBP structures

In the following we repeat the experiments from section 3 but also test networks with added RBP structures. For convenience we repeat the previous results in the tables in this section.

5.1 Classification experiments

This experiment is analogous to the first classification experiment. In the case of the feed-forward network, RBP2(a) was used in the mid fusion approach and for recurrent network, RBP2(b) was used. We trained again for 10 epochs and all networks converged to perfect classification on the training set. Table 4 provides the overall test accuracy for the three approaches.

Task	RBP	FFNN	RNN	GRU	LSTM
1a) ABA vs other	-	50%	55%	55%	55%
	RBP1n	50%	55%	55%	55%
	RBP1p	65%	70%	70%	70%
	RBP2	100%	100%	100%	100%
1b) ABB vs other	-	50%	55%	55%	55%
	RBP1n	50%	55%	55%	55%
	RBP1p	65%	70%	70%	70%
	RBP2	100%	100%	100%	100%
2) ABA vs ABB	-	50%	50%	50%	50%
	RBP1n	50%	60%	65%	65%
	RBP1p	75%	75%	75%	75%
	RBP2	100%	100%	100%	100%
3) ABC vs other	-	50%	50%	50%	50%
	RBP1n	55%	65%	65%	65%
	RBP1p	55%	70%	70%	70%
	RBP2	100%	100%	100%	100%
4) ABA-BAB vs other	-	50%	50%	50%	50%
	RBP1n	55%	72%	75%	75%
	RBP1p	69%	74%	75%	76%
	RBP2	100%	100%	100%	100%

Table 4: Classification experiments with RBP: test accuracy for the different models and tasks, as explained above. Results with '-' in the RBP column are the same as in section 3 and shown here again for comparison.

The results with RBP1n models already show some improvement over the baseline in most configurations, but the result are only slightly above the standard networks, with RNNs, GRUs and LSTMs benefiting more than FFNN. This supports our hypothesis that learning to compare corresponding input neurons is a challenging task for neural networks. However, the results show that providing that comparison is not sufficient for learning identity rules.

RBP1p structures also aggregate all the $l\ DR_n$ neurons that belong to a pair of input tokens. The results show that providing that information leads to improved accuracy and provide evidence that this aggregation is another necessary step that the networks do not learn reliably from the data.

The RBP2 models enable the neural networks to make predictions and classifications that generalise according to identity rules that it learns from data. The RBP2 leads to perfect classification for all network types tested. This confirms the design consideration that comparing pairs of tokens provides the relevant information in the form required for classification, as the classes are defined by *equals* relations, so that the activations of the DRp units are directly correlated with the class labels.

A surprising result is the big difference between the generalisation using the RBP1p and the RBP2 structures. They both provide the same information, only in different layers of the network, but RBP1p only reaches at most 75% with a 50% baseline. We hypothesize that the additional expressive power provided by the non-linearities in the hidden layer here hinders effective learning. This effect deserves further investigation.

5.2   Prediction experiments

Here we performed two experiments separately on ABA and ABB patterns as in experiment 1 on prediction. The tasks are the same as previously and we trained again for 10 epochs after which all networks had converged to perfect prediction accuracy on the training data. Table 5, summarises the accuracy for RNN, GRU and LSTM without RBP, and with RBP1n, 1p, 2, and 3.

Overall, we observe that only the LSTM benefits from RPB1n, RBP1p, and RBP2 structures, all other networks can apparently not make use of the information provided. The RBP3 model, on the other hand, leads to perfect classification on our synthetic dataset.

Our interpretation is that standard recurrent networks do not learn the more complex mapping that prediction requires, as not only recognition of a pattern but also selecting a prediction on the basis of that pattern is required. The somewhat better results of the LSTM networks are interesting. In the RBP3 model, the mapping between the identity patterns and back to the vocabulary adds considerable prior structure to the model and it is very effective in achieving the generalisation of rule-based patterns.

Pattern	RBP	RNN	GRU	LSTM
1) ABA	-	0%	0%	0%
	RBP1n	0%	0%	16%
	RBP1p	0%	0%	18%
	RBP2	0%	0%	20%
	RBP3	100%	100%	100%
2) ABB	-	0%	0%	0%
	RBP1n	0%	0%	17%
	RBP1p	0%	0%	20%
	RBP2	0%	0%	22%
	RBP3	100%	100%	100%

Table 5: Test set accuracy in prediction experiments for patterns ABA and ABB. As before, results are averaged over 10 simulations and rounded to the nearest decimal point. Results with '-' in the RBP column are the same as in section 3 and shown here again for comparison.

6 Experiment 3: mixed tasks and real data

The results presented here were all obtained with synthetic data where classification was exclusively on rule-based abstract patterns. This raises the question whether the RBP will impede recognition of concrete patterns in a mixed situation. Furthermore, we would like to know whether RBP is effective with real data where the abstract and concrete patterns may interact.

6.1 Mixed abstract and concrete patterns

We conducted an experiment where the classes were defined by combinations of abstract and concrete patterns. Specifically we defined 4 classes based on the abstract patterns ABA and ABB combined with the concrete patterns $a**$ and $b**$. E.g., the class $ABA, a**$ can be expressed logically as

$$eq(\alpha, \gamma) \wedge \neg eq(\alpha, \beta) \wedge \alpha = \text{`}a\text{'}. \tag{5}$$

We use a vocabulary of 18 characters, out of which 12 are used for training and 6 are used for validation/testing in addition to 'a' and 'b', which need to appear in all sets because of the definition of the concrete patterns. For class 1/3 and class 2/4, abstract patterns ABA and ABB are used respectively. Class 1/2 and 3/4 start with tokens 'a' and 'b' respectively. The train, validation and test split is 50%, 25%, and 25% respectively. We trained the network for 10 epochs, leading to perfect classification on the training set. A total of 10 simulations has been performed. We test a feed forward and a recurrent neural network without and with RBP1p and RBP2. The results are shown in Table 6.

RBP	FFNN	RNN
-	23%	42%
RBP1p	49%	57%
RBP2	100%	100%

Table 6: Test set accuracy for mixed abstract/concrete pattern classification.

As in the previous experiments, networks without RBP fail to generalise the abstract patterns. The results for RBP1p and RBP2 show, that the ability to learn and recognise the concrete patterns is not impeded by adding the RBP structures.

6.2 Language models with RBP

In order to test the capability of networks with RBP structure, we use them in two language modelling tasks. One is to predict characters in English text, and one is to predict the pitch of the next note in folk song melodies. We selected both tasks because of the prevalence of repetitions in the data, as notes in music and characters in English tend to be repeated more than words. Our RBP structures are designed to model identity-rules and we therefore expect them to be more effective on tasks with more repetitions.

Character prediction We conducted a character prediction experiment on a subset of the Gutenberg electronic book collection[2], consisting of text with the dataset size of 42252 words. We used 2 hidden layers with 50 neurons each. In the RBP2 model, the DRp units were concatenated with the first hidden layer. The learning rate is set to 0.01 and the network training converged after 30 epochs. Each character is predicted without and with the RBP variants using a context size of 5. The prediction results are summarized in Table 7.

Pitch prediction In another experiment we applied RBP to pitch prediction in melodies [42] taken from the Essen Folk Song Collection [43]. We performed a grid search for each context length for hyper parameter tuning, with [10,30,50,100] as the size of the hidden layer and [30,50] epochs with learning rate set to 0.01 with one hidden layer. The results for context length 5 are summarized in Table 8. RBP

---

[2]https://www.gutenberg.org/

RBP	RNN	GRU	LSTM
-	3.8281	3.8251	3.8211
RBP1p	4.4383	4.4368	4.4321
RBP2	3.7512	3.7463	3.7448
RBP3	**3.4076**	**3.4034**	**3.4012**

Table 7: Character prediction task. The numbers show the average cross entropy loss per character on the test set (lower is better, best values are set in bold), without and with RBP structures using context length 5.

improved the network performance for RNN, GRU, and LSTM. Overall, LSTM with late fusion produces the best result and also improves over the best reported performance in pitch prediction with a long-term model on this dataset with a cross-entropy of 2.547, which was achieved with a feature discovery approach by [44].

RBP	RNN	GRU	LSTM
-	2.6994	2.5702	2.5589
RBP1p	2.6992	2.5714	2.5584
RBP2	2.6837	2.5623	2.5483
RBP3	**2.6588**	**2.5549**	**2.5242**

Table 8: Pitch prediction task on the Essen Folk Song Collection. The numbers show the average cross entropy per note (lower is better, best values are set in bold), without and with RBP using context length 5.

Results In both character and pitch prediction, the addition of RBP3 structures improves the overall results consistently. RBP1n leads to a deterioration in character prediction and to inconsistent effect on pitch prediction, while RBP2 leads to a slight but consistent improvement in both tasks. This provides further evidence that the RBP structure enables the learning of relevant patterns in the data.

7 Discussion

7.1 Standard neural networks

The results of the experiments described above confirm the results of [1] and others that standard recurrent (and feed-forward) neural networks do not learn generalisable identity rules. From the tested models and settings of the task we can see that the lack of activation of input neurons impedes learning, but avoiding this lack is not sufficient. The task assumes that the identity of input tokens is easy to recognise, classify and base predictions on, but the models we tested do not learn to generalise in this way. These results confirm our view that in order to generalise it is necessary to know which input neurons are related, similarly on the next level, which comparisons of input belong to a pair of tokens so that they can be aggregated per token. The structure of neural networks does not provide any prior preference for inputs that are related in this way over any other combinations of inputs. This makes it seem plausible that the solutions by [5, 6, 9] could not be replicated by [7, 10].

7.2   Constructive model with RBP

The RBP model addresses the learning of identity rules by adding neurons and connections with fixed weights. From the input neurons we add connections to a DR (differentiator-rectifier) unit from each pair of corresponding input neurons within any pair of tokens (represented in one-hot encoding). These DR units calculate the absolute of the difference of the activations of the two input neurons. They are followed by DRp units that aggregate by taking the sum of the DR unit activations for each pair of tokens. The fact that the DRp units relate to the difference between each pair of neurons makes the learning task for classification much simpler, as has been confirmed by our results. An open question in this context is why the RBP2 is so much more effective than RBP1p for classification, although the only difference is the layer in which the information is added into the network.

For prediction, we need a more complex structure, as beyond recognition of identity, also the selection of the token to predict is required, that depends on the tokens in the context and their similarity relations. The constructive RBP solution requires a transformation into a representation of identity relations in the input that is mapped to identities between input and output and that is mapped back to the token space by adding prediction probability to the tokens that are predicted to be identical between input and output. This created a complex predefined structure, but without it even the models that achieved prefect classification failed to make correct predictions with new data. Only the LSTM models could use the RPB1 and RBP2 information to make prediction above the baseline (22% vs 8.3%). We hypothesise that the gating structure of the LSTMs enables at least some effective mapping. The 100% correct predictions by all models using RBP3 shows the effectiveness of this structure.

7.3 Applications

Adding a bias into the network with a predefined structure such as RBP raises the question whether there is a negative effect on other learning abilities of the network and whether interactions between the abstract and concrete tasks can be learnt. In the mixed pattern experiment, RBP is still effective and showed no negative effect. In experiments with real language and music data we found that RBP3 has a positive effect on the prediction of characters in language and pitches in folk song melodies. The small negative effect of RBP1 on character prediction seems to indicate that there may be confounding effect where identity rules are less relevant. This effect did not appear in melody prediction, where repetition is more important.

7.4 Extrapolation and inductive bias

The results in this study confirm that an inductive bias is needed for extrapolation, in the terminology of [28], in order to generalise in some sense outside the space covered by the training data. This general challenge has recently attracted some attention. E.g., [45] provided several solutions to the related problem of learning equality of numbers (in binary representation), which does not generalise from even to odd numbers as pointed out already by [46]. As the authors point out in [45], an essential question is which biases are relevant to the domain and problem. The identity problem addressed here is in itself fundamental to learning about relations [47], as relations depend on object identity. This further raises the question what is needed to enable more complex concepts and rules to be learnt, such as more general logical concepts and rules.

The identity rules also point to the lower-level problem that the natural relations of position and belonging to objects are not naturally addressed in neural networks. Other tasks may require different structures, relating for example to arithmetics, geometry or physics [48]. We therefore see as an important task the definition or predefined structures in neural networks, so that they create useful inductive bias, but do not prevent learning of functions that do not conform to that bias.

8   Conclusions

Our experiments show that the observation by [1], that neural networks are unable to learn general identity rules, holds for standard feed-forward networks, recurrent neural networks, and networks of GRUs and LSTMs. The solution we propose here, the Relation Based Patterns (RBP), introduce an additional structure with fixed weights into the network. Our experiments confirm that the RBP structures enable the learning of abstract patterns based on identity rules in classification and prediction as well as in mixed abstract and concrete patterns. We have further found that adding RBP structures improves performance in language and music prediction tasks.

Overall, we find that standard neural networks do not learn identity rules and that adding RBP structure creates an inductive bias which enables this extrapolation beyond training data with neural networks. This outcome raises the question on how to develop further inductive biases for neural networks to improve generalisation of learning on other tasks and more generally.

References

[1] G. F. Marcus, S. Vijayan, S. Rao, P. Vishton, Rule learning by seven-month-old infants, Science, 283 283 (5398) (1999) 77–80.

[2] P. C. Gordon, K. J. Holyoak, Implicit learning and generalization of the mere exposure effect., Journal of Personality and Social Psychology, 45, 492–500.

[3] B. J. Knowlton, L. R. Squire, The information acquired during artificial grammar learning., Journal of Experimental Psychology: Learning, Memory, and Cognition, 20, 79 –91.

[4] B. J. Knowlton, L. R. Squire, Artificial grammar learning depends on implicit acquisition of both abstract and exemplar-specific information, Journal of Experimental Psychology: Learning, Memory, and Cognition, 22, 169 –181.

[5] M. Seidenberg, J. Elman, Do infants learn grammar with algebra or statistics?, Science 284 (5413) (1999) 433–433.

[6] J. Elman, Generalization, rules, and neural networks: A simulation of Marcus et. al, https://crl.ucsd.edu/~elman/Papers/MVRVsimulation.html.

[7] M. Vilcu, R. F. Hadley, Generalization in simple recurrent netrworks, in: Proceedings of the Annual Meeting of the Cognitive Science Society, Vol. 23, 2001, pp. 1072–1077.

[8] T. R. Shultz, A. C. Bale, Neural network simulation of infant familiarization to artificial sentences: Rule-like behavior without explicit rules and variables, Infancy, 2:4, 501-536, DOI: 10.1207/S15327078IN020407.

[9] G. Altmann, Z. Dienes, Rule learning by seven-month-old infants and neural networks, In Science 284 (5416) (1999) 875–875.

[10] M. Vilcu, R. F. Hadley, Two apparent 'counterexamples' to Marcus: A closer look, Minds and Machines 15 (3-4) (2005) 359–382.

[11] T. R. Shultz, J.-P. Thivierge, D. Titone, Generalization in a model of infant sensitivity to syntactic variation, in: Proceedings of the Annual Meeting of the Cognitive Science Society, 2005, pp. 2009–20014.

[12] L. Shastri, S. Chang, A spatiotemporal connectionist model of algebraic rule-learning, Tech. Rep. TR-99-011, Berkeley, California: International Computer Science Institute

(1999).

[13] M. Gasser, E. Colunga, Babies, variables, and connectionist networks, in: Proceedings of the 21st Annual Conference of the Cognitive Science Society, Lawrence Erlbaum, 1999, p. 794.

[14] P. F. Dominey, F. Ramus, Neural network processing of natural language: I. sensitivity to serial, temporal and abstract structure of language in the infant, Language and Cognitive Processes 15 (1) (2000) 87–127.

[15] M. H. Christiansen, C. M. Conway, S. Curtin, A connectionist single-mechanism account of rule-like behavior in infancy, in: Proceedings of the 22nd annual conference of the cognitive science society, 2000, pp. 83–88.

[16] R. G. Alhama, W. Zuidema, Pre-wiring and pre-training: What does a neural network need to learn truly general identity rules, CoCo at NIPS.

[17] J. A. Fodor, Z. W. Pylyshyn, Connectionism and cognitive architecture: A critical analysis, Cognition 28 (1-2) (1988) 3–71.

[18] R. F. Hadley, Systematicity in connectionist language learning, Mind & Language 9 (3) (1994) 247–272.

[19] P. Smolensky, The constituent structure of connectionist mental states: A reply to Fodor and Pylyshyn, Southern Journal of Philosophy 26 (Supplement) (1987) 137–161.

[20] J. Fodor, B. P. McLaughlin, Connectionism and the problem of systematicity: Why Smolensky's solution doesn't work, Cognition 35 (2) (1990) 183–204.

[21] D. Chalmers, Why Fodor and Pylyshyn were wrong: The simplest refutation, in: Proceedings of the Twelfth Annual Conference of the Cognitive Science Society, Cambridge, Mass, 1990, pp. 340–347.

[22] L. Niklasson, T. van Gelder, Systematicity and connectionist language learning, Mind and Language 9 (3) (1994) 28–302.

[23] M. H. Christiansen, N. Chater, Generalization and connectionist language learning, Mind & Language 9 (3) (1994) 273–287.

[24] R. F. Hadley, Systematicity revisited: reply to Christiansen and Chater and Niklasson and van Gelder, Mind & Language 9 (4) (1994) 431–444.

[25] S. L. Frank, Getting real about systematicity, in: P. Calvo, J. Symons (Eds.), The architecture of cognition: Rethinking Fodor and Pylyshyn's systematicity challenge, MIT Press, 2014, pp. 147–164.

[26] B. Lake, M. Baroni, Generalization without systematicity: On the compositional skills of sequence-to-sequence recurrent networks, in: International Conference on Machine Learning, 2018, pp. 2879–2888.

[27] I. Sutskever, O. Vinyals, Q. V. Le, Sequence to sequence learning with neural networks, in: Advances in neural information processing systems, 2014, pp. 3104–3112.

[28] G. F. Marcus, Deep learning : a critical appraisal, arXiv:1801.00631.

[29] S. Sabour, N. Frosst, G. E. Hinton, Dynamic routing between capsules, in: Advances in Neural Information Processing Systems, 2017, pp. 3856–3866.

[30] K. Kansky, T. Silver, D. A. Mély, M. Eldawy, M. Lázaro-Gredilla, X. Lou, N. Dorfman,

S. Sidor, S. Phoenix, D. George, Schema networks: Zero-shot transfer with a generative causal model of intuitive physics, arXiv preprint arXiv:1706.04317.

[31] C. Szegedy, W. Zaremba, I. Sutskever, J. Bruna, D. Erhan, I. J. Goodfellow, R. Fergus, Intriguing properties of neural networks, CoRR abs/1312.6199. arXiv:1312.6199. URL http://arxiv.org/abs/1312.6199

[32] R. Feinman, B. M. Lake, Learning inductive biases with simple neural networks (2018). arXiv:1802.02745.

[33] D. G. T. Barrett, F. Hill, A. Santoro, A. S. Morcos, T. Lillicrap, Measuring abstract reasoning in neural networks (2018). arXiv:1807.04225.

[34] Y. Bengio, A. Courville, P. Vincent, Representation learning: A review and new perspectives, IEEE transactions on pattern analysis and machine intelligence 35 (8) (2013) 1798–1828.

[35] D. E. Rumelhart, G. E. Hinton, R. J. Williams, Learning internal representations by error propagation, Tech. rep., California Univ San Diego La Jolla Inst for Cognitive Science (1985).

[36] J. L. Elman, Finding structure in time, Cognitive science 14 (2) (1990) 179–211.

[37] K. Cho, B. Van Merriënboer, C. Gulcehre, D. Bahdanau, F. Bougares, H. Schwenk, Y. Bengio, Learning phrase representations using rnn encoder-decoder for statistical machine translation, arXiv preprint arXiv:1406.1078.

[38] S. Hochreiter, J. Schmidhuber, Long short-term memory, Neural Computation 9 (8) (1997) 1735–1780. doi:10.1162/neco.1997.9.8.1735. URL https://doi.org/10.1162/neco.1997.9.8.1735

[39] D. P. Kingma, J. Ba, Adam: A method for stochastic optimization, arXiv:1412.6980.

[40] M. Leshno, V. Y. Lin, A. Pinkus, S. Schocken, Multilayer feedforward networks with a nonpolynomial activation function can approximate any function, Neural networks 6 (6) (1993) 861–867.

[41] H. Siegelmann, E. Sontag, On the computational power of neural nets, Journal of Computer and System Sciences (1995) Volume 50, Issue 1, pp. 132–150.

[42] R. M. Kopparti, T. Weyde, Evaluating repetition based melody prediction over different context lengths, ICML Joint Workshop on Machine Learning and Music, Stockholm, Sweden, July 14.

[43] H. Schaffrath, The essen folksong collection in the humdrum kern format, Menlo Park, CA, Centre for Computer Assissted Research in the Humanities, 1995.

[44] J. Langhabel, R. Lieck, M. Rohrmeier, Feature discovery for sequential prediction of monophonic music, International Society for Music Information Retrieval Conference (2017) 649–655.

[45] J. Mitchell, P. Minervini, P. Stenetorp, S. Riedel, Extrapolation in NLP, arXiv:1805.06648.

[46] G. F. Marcus, The algebraic mind: Integrating connectionism and cognitive science, Cambridge MIT Press.

[47] P. W. Battaglia, J. B. Hamrick, V. Bapst, A. Sanchez-Gonzalez, V. Zambaldi, M. Ma-

linowski, A. Tacchetti, D. Raposo, A. Santoro, R. Faulkner, et al., Relational inductive biases, deep learning, and graph networks, arXiv preprint arXiv:1806.01261.

[48] N. Cohen, A. Shashua, Inductive bias of deep convolutional networks through pooling geometry (2016). `arXiv:1605.06743`.

Received 21 June 2018

www.ingramcontent.com/pod-product-compliance
Lightning Source LLC
Chambersburg PA
CBHW062026210326
41519CB00060B/7136